24 SILICON-CONTROLLED RECTIFIER PROJECTS

Other TAB Books by the Author

No. 1316 *The Master Handbook of Telephones*
No. 1419 *Astronomy & Telescopes: A Beginner's Handbook (coauthored with Susan E. Harris)*
No. 1433 *Make Your Own Professional Home Video Recordings*
No. 1487 *The Shortwave Listener's Antenna Handbook*
No. 1492 *39 One-Evening Electronic Projects*
No. 1496 *The IBM Personal Computer*
No. 1526 *33 Games of Skill and Chance for the IBM PC®*
No. 1536 *Beginner's Guide to Reading Schematics*
No. 1556 *Graphics Programs for the IBM PC®*
No. 1596 *Music and Speech Programs for the IBM PC®*
No. 1909 *Build a Personal Earth Station for Worldwide Satellite TV Reception—2nd Edition*
No. 1956 *The Quadram Connection*
No. 1996 *The 555 IC Project Book*
No. 2097 *All About Electric & Hybrid Cars*
No. 2321 *The Joy of Flying*

870320

No. 1586
$17.95

24 SILICON-CONTROLLED RECTIFIER PROJECTS

ROBERT J. TRAISTER

TAB BOOKS Inc.
Blue Ridge Summit, PA 17214

To my fishing friend, Jerry Matlick, an individual whose self control is beyond reproach.

FIRST EDITION

FIRST PRINTING

Printed in the United States of America

Library of Congress Cataloging in Publication Data

Traister, Robert J.
24 silicon-controlled rectifier projects.

Includes index.
1. Silicon-controlled rectifiers. I. Title.
II. Title: Twenty-four silicon-controlled rectifier projects.
TK7872.R35T74 1986 621.31'37 85-27778
ISBN 0-8306-0386-7
ISBN 0-8306-0486-3 (pbk.)

Front cover photograph courtesy of Motorola Semiconductor Products Inc.

Contents

Introduction

THE THYRISTOR IS NOT SO MUCH A DEVICE IN ITSELF AS A WHOLE family of power-controlling solid-state components. They are simple, yet they perform highly complex functions not possible a few decades ago. The thyristor family is made up of silicon-controlled rectifiers, triacs, diacs, asymmetrical rectifiers, and other components that can be used as solid-state switches, for variable power control, and for low-voltage control of high-voltage/high-power circuits.

Can you remember the days when a large variable resistor was used to control power to an electronic load? This bulky, extremely expensive device has been replaced with a thyristor that may have cost only a couple of dollars. The thyristor is quite small and can easily be held in the palm of the hand. Yet it does everything the variable resistor did and much more efficiently, with a negligible loss of power. The light dimmer controls in some homes are excellent examples of what I'm talking about. Before the thyristor, these were heavy variable resistors. Now they are miniature devices that do not even fill the space of a normal light switch.

This is only a small sampling of what thyristors can accomplish. From a hobby standpoint, silicon-controlled rectifiers and triacs are an integral part of such circuits as light modulators, or light organs as they're sometimes called. The diminutive signal from a stereo system is used to control the power to quite large lighting systems. Thyristors may be used as motor speed controls and even in highly efficient electronic filtering circuits for dc power supplies. These components bridge the gap between the hobbyist and the electronic technician, being used in devices for the home as well as in extremely complex scientific equipment.

Aside from their cost effectiveness, thyristors help to make power circuits quite simple. For a few dollars, you can now perform control functions that cost thousands or even hundreds of thousands of dollars a few decades ago.

The circuits, discussions, and projects included in this book were specially chosen to provide a learning experience for the beginner and a challenge to the experienced builder. Combining theory with building practical circuits will yield the most efficient results from this text.

Chapter 1

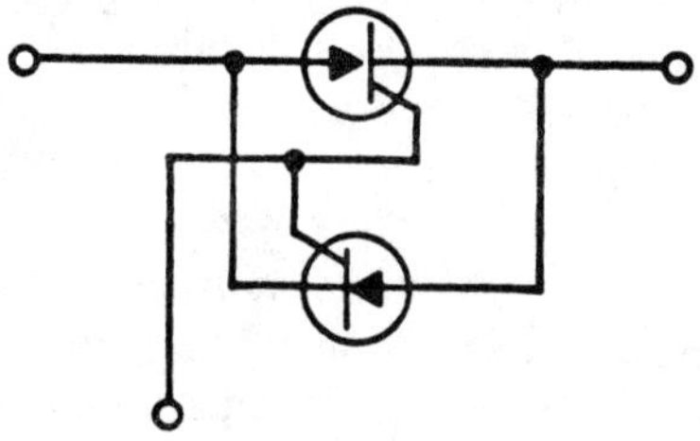

Silicon-Controlled Rectifiers and Other Thyristors

THE YEAR 1982 MARKED THE SILVER ANNIVERSARY OF THE DEVELOPment of the first commercially practical silicon-controlled rectifier (SCR) by General Electric Company. It is hard to believe that that first, small, 35 A, 200 V device would trigger a revolution in electric power control and conversion. But today it would be even more difficult to imagine power control equipment without the SCR, since it has proliferated in so many different types, sizes, ratings and applications. SCRs have grown smaller: One hundred SCRs rated at 200 V, 0.8 A control teletypewriters. They have grown bigger; 4500 V, 3000 A brutes transmit megawatts in HVDC systems. They have gone out into space and deep into the earth. They have been made faster to operate at many kilohertz in induction heating equipment. Other typical applications include ac and dc motor speed controls, lighting and temperature controls, welding, uninterruptible power supplies, electromechanical refining, electrostatic precipitators, fusion and laser power supplies, traction and pulse modulators. The list goes on and on.

Because General Electric Company (GE) was a pioneer in the development of the silicon-controlled rectifier, I went to them for research information about SCRs. They responded with a wealth of theoretical and practical application information, much of which is included throughout this book.

While the SCR remains the "general practitioner" of the industry, it has been joined by many other special types of thyristors developed to fulfill different circuit needs. In 1962, GE announced another first, the triac, which has found wide acceptance in consumer and light industrial controls. This was followed in 1963 by the reverse conducting SCR (an SCR in antiparallel with a diode). The amplifying gate structure, first developed by GE in 1965 to increase SCR di/dt ratings, is now standard on all high current SCRs to keep gating current requirements within reasonable limits. The asymmetric SCR is a more efficient inverter type that has evolved as a result of technological innovations. The gate turn-off SCR (GTO) negates the requirement for commutation circuitry, and the reverse conducting thyristor incorporates an antiparallel diode in its structure to conduct reactive currents when driving inductive loads.

By solid state electronic standards, a 25-year old product is considered mature, if not obsolete. This is not true of the silicon-controlled rectifier. To the contrary, significant development work continues to improve the dynamic performance of SCRs and to invent new types for specialized needs. New packaging concepts have emerged as better silicon passivation techniques allow the silicon subassemblies to be encapsulated in non-hermetic housings or even to be immersed directly in a cooling medium, such as transformer oil. Engineered plastic housings not only reduce manufacturing costs, but can also reduce equipment assembly costs by making possible new multiple device modules that are electrically isolated from their mounting base plate. Light-triggered SCRs, field-controlled, and gate-assisted turn-off SCRs are being pursued, while new equipment designs demand higher voltage and higher current SCRs.

In order to understand the operation of the silicon-controlled rectifier and other similar devices, it is necessary to first learn about the makeup and operation of simple pn junction rectifiers. This is the simplest form of practical semiconductor device while simultaneously being a fundamental building block for all other bipolar structures such as npn and pnp transistors, thyristors, and related products. A knowledge of the theoretical behavior of a rectifier is therefore vital to an understanding of the operation of more complex devices.

A pn rectifier is the juxtaposition of two dissimilar layers of silicon that have been treated (doped) to provide, in one case, a surfeit of mobile electrons (n-material) and, in the other case, a surfeit of mobile positive carriers (holes, or p-material). When these two materials are placed side by side, as in a pn junction, the natural affinity of dissimilar charges that meet and cancel each other out creates a region at the meeting plane devoid of all free electrical charge; this region is characterized by a high resistance to the flow of electric current (no carriers). To force current conduction, it is necessary to apply a voltage (positive to anode-p) such that the electric field resulting from this voltage overcomes the natural barrier and permits the flow of current from anode to cathode. It is this phenomenon that explains the 0.7-volt threshold voltage associated with forward conduction across a silicon pn junction.

When the external voltage is reversed in polarity, however, the natural space charge region is expanded instead of being opposed, making the reverse conduction even more difficult. This phenomenon explains the absence of significant reverse current across a reverse-biased pn junction. Despite the repulsion of like charges resulting from reverse bias, a few carriers of opposite polarity on either side of the junction are accelerated across the junction by the presence of this same electric field. As the magnitude (volts/meter) of the electric field is increased, these "leakage" carriers gain additional kinetic energy, so that as they traverse the junction region, they are able, when they inevitably collide with fixed atoms in the crystalline structure, to liberate additional carriers that supplement the original leakage current. These additional carriers, in turn, are accelerated, and create in their turn more current by subsequent collisions. This phenomenon is known as *reverse avalanche* or *reverse breakdown*. The electrical characterization of a pn junction under both forward and reverse bias is shown in Fig. 1-1.

When in the forward biased mode with significant current flow, the junction cannot instantaneously revert to the blocking mode just described because the junction zone is replete with charges already in motion. These charges must first be swept out by current reversal or by recombination (positive charges being neutralized by meeting negative charges). The time this reverse current flows for is called *reverse recovery time*, and the charge swept out is called *reverse recovery charge*. High frequency circuits need rectifiers in which reverse recovery time is held to a minimum by special fabrication techniques, such as gold doping or electron irradiation, that encourage rapid recombination of the carriers crossing the junction region.

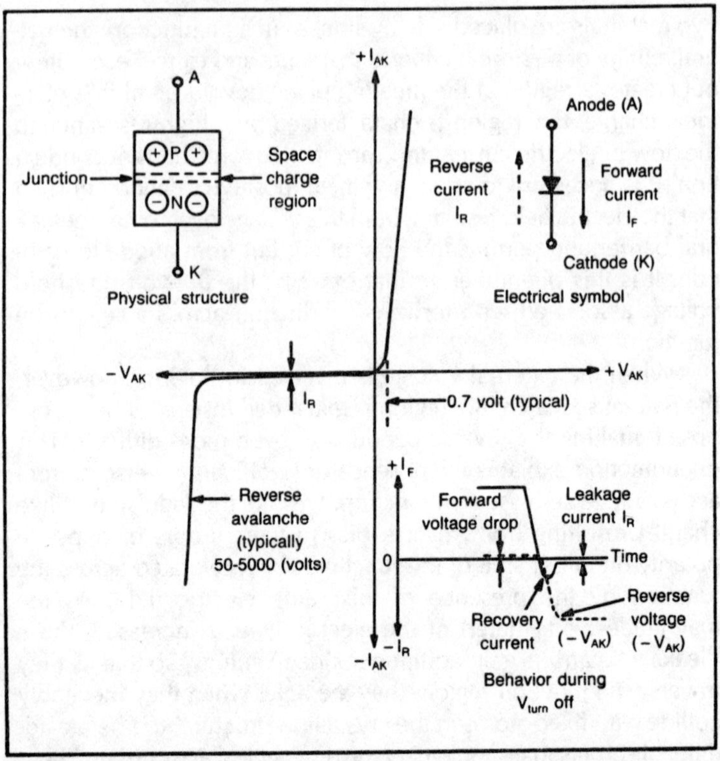

Fig. 1-1. Electrical characteristics of a PN junction diode (courtesy General Electric).

THYRISTORS

A *thyristor* is any semiconductor switch whose bistable action depends on pnpn regenerative feedback. Thyristors can be two-, three-, or four-terminal devices, and both unidirectional and bidirectional devices are available.

The silicon-controlled rectifier (SCR) is by far the best known member of the thyristor family. Because it is a three-terminal device (anode, cathode, and control gate) that can conduct current in one direction only (anode to cathode), the SCR is classified as a reverse blocking triode thyristor. Some other reverse blocking triode thyristors are the silicon unilateral switch (SUS), the light-activated silicon-controlled rectifier (LASCR), the complementary SCR (CSCR), the gate turn-off switch (GTO) and the programmable unijunction transistor (PUT). A special case is the asymmetrical SCR (ASCR) which, unlike conventional SCRs that display approximately equal forward and reverse blocking capabilities, is deliberately constructed with a very low reverse blocking voltage. The silicon-controlled switch (SCS) is a reverse blocking tetrode thyristor (it has two control gates referenced respectively to anode and cathode), while the Shockley diode is a reverse blocking diode thyristor (triggered by anode voltage).

Bidirectional thyristors are categorized as pnpn devices that can conduct in either direction. Typical examples include the triac (for triode ac switch) and the silicon bilateral switch (SBS). Once again, a special case is the reverse conducting thyristor (RCT), which consists of an asymmetrical SCR and companion bypass diode (pn junction) integrated into a single chip. Figure 1-2 depicts the symbols and basic electrical characteristics of these various thyristor types.

Since General Electric first introduced the SCR in 1957, circuit designers have been seeking ever higher levels of device performance. Because thyristors are used almost exclusively as power switches, performance can be related to the ability of the device to switch efficiently high current loads from high source voltages with maximum speed. The parameters associated with these criteria are:

- Blocking voltage.
- Current carrying capability.
- Di/dt (permissible rate of rise of anode current at turn-on).
- Dv/dt (allowable rate of application of forward blocking voltage).
- Turn-off time.
- Gating requirements.
- Maximum operating frequency.

Various techniques to improve device performance are available. However, as the following discussion will show, many of these techniques lead to performance trade-offs elsewhere.

Voltage

Surface contouring, or beveling, allows higher voltage operation by reducing the electric fields at the junction periphery, the critical interface between a blocking junction and its external environment. Reduced electric fields here permit higher voltage operation before breakdown occurs. Beveling is advantageous because only a slight reduction in usable cathode area is sacrificed.

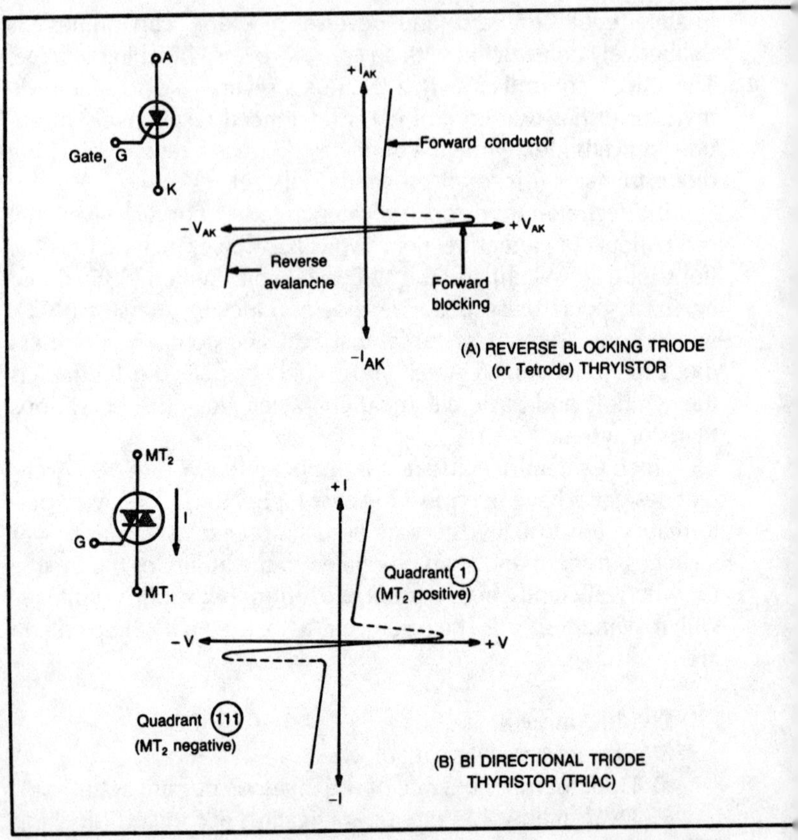

Fig. 1-2. Symbols and characteristics of various thyristors (courtesy General Electric).

Beveling techniques are usually limited to premium-grade industrial thyristors only, where the added cost of round pellets can be justified. Beveling effectiveness is seriously affected by field concentration at the corners when the technique is applied to the rectangular pellets usually prescribed for consumer-oriented products. Other techniques, such as modifying base width or silicon resistivity, or increasing minority carrier lifetime, while effective, do result in performance sacrifices elsewhere.

Current

The allowable current through a device depends primarily upon the on-state voltage. Any variable that decreases this voltage drop (and hence, internal power dissipation), raises the cur-

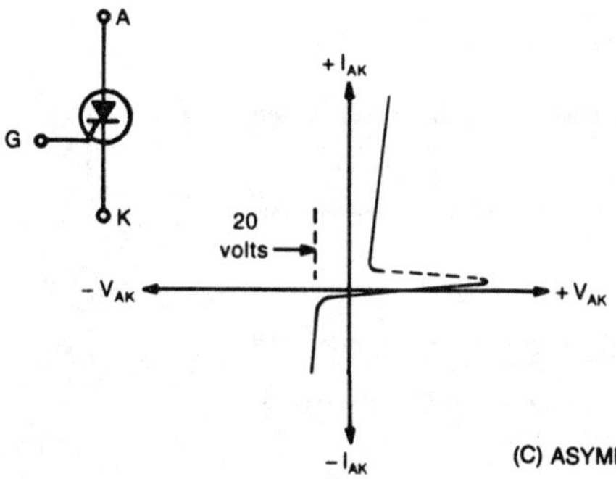

(C) ASYMMETRICAL SCR

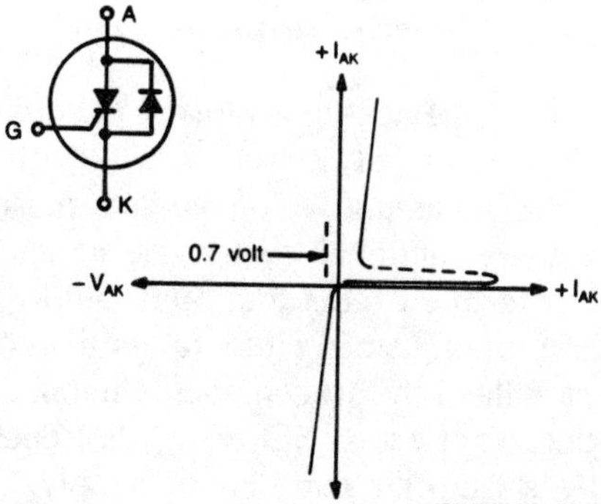

(D) REVERSE CONDUCTING THYRISTOR (RCT)

rent limit. Favorable effects are larger pellets, small B_N base width, higher lifetimes, and lower silicon resistivity.

Di/dt

Di/dt failures are essentially precipitated by too high a concentration of current in too small a pellet area during initial turn-on of the thyristor, with the resultant high temperature gradient leading to a localized melting. The obvious solution is to turn on more area at the instant of gate firing, and this is precisely the objective of the structures to be shown. As a starting point, Fig. 1-3 shows the traditional side-gate, or point-gate, featured in the earliest designs and still used today in some low-cost consumer and light industry-oriented products. This is the simplest gate structure,

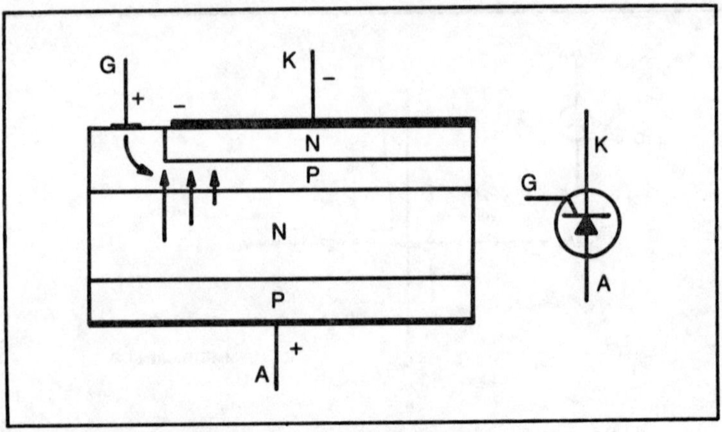

Fig. 1-3. Conventional side gate SCR (courtesy General Electric).

perfectly adequate for small area devices or for low di/dt applications of larger thyristors. The area initially turned on is quite small and is heavily dependent on the amplitude of the gate pulse.

Figure 1-4 illustrates a conventional center-gate structure, similar in function to the side-gate except that di/dt ratings are generally higher due to turn on of a small circle rather than of a point. Figure 1-5 portrays a state-of-the-art center-amplifying gate structure that initiates conduction of a small SCR. The gating energy in this latter phase is derived from the main anode supply. The

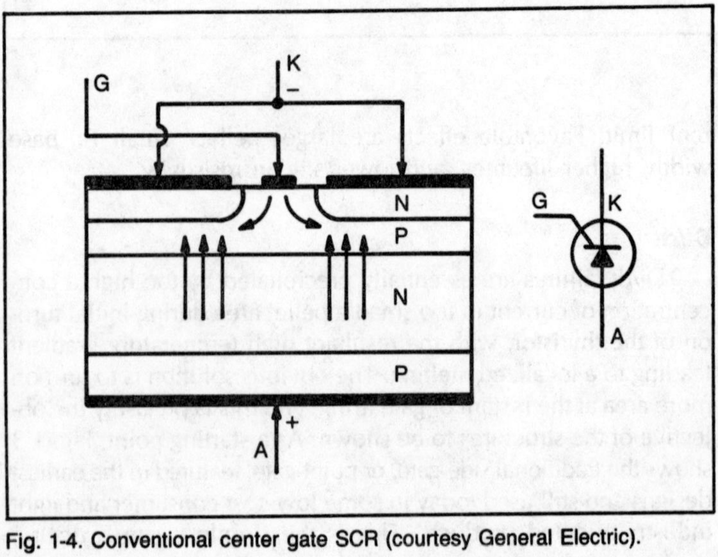

Fig. 1-4. Conventional center gate SCR (courtesy General Electric).

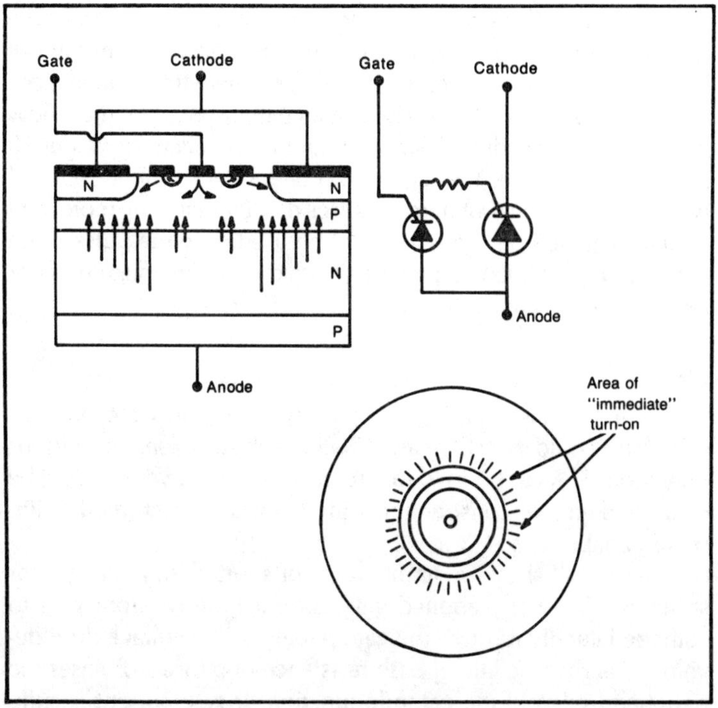

Fig. 1-5. Amplifying gate SCR (courtesy General Electric).

double switching sequence is optimized by choice of proper geometry such that the principal thyristor switches on immediately after the pilot is activated. The area of immediate turn-on is sizable as shown on the plan view of the pellet structure in Fig. 1-5.

Although it is possible to turn on a ring around the amplifying gate as previously described, a finite spreading velocity requires a finite period of time before conduction moves out from the periphery of the amplifying gate to cause the entire annular portion of the thyristor to be on. Depending on the designed voltage and speed characteristics of a thyristor, the spreading velocity can range anywhere from 3000 to 8000 cm/second. This represents spreading times of 50 μs for a 110 A device to better than 300 μs for a 550 A device. One can appreciate that for narrow pulses (less than the above times), something less than the entire pellet is in conduction and wide pulse current capability cannot be realized.

The solution employed to extend the amplifying gate principle to thyristors designed for narrow pulse application is interdigitation. Figure 1-6 shows the extension of the amplifying gate technology to include interdigitation. The fingers extending out into the cathode area increase the gate periphery from a little more than 2 cm to about 44 cm. The distance that spreading must now traverse is much reduced, so that total turn-on occurs in about 10 to 30 μs. Of course, the region for conduction is much reduced with interdigitation, but narrow pulses don't require any more, and an interdigitated device would not be used for wide pulses. The previously mentioned non-interdigitated devices are much superior for wide pulse applications.

Dv/dt

As mentioned earlier, a rapidly rising forward voltage waveform could switch on an SCR. Since this can lead to spurious operation, SCRs to be used in circuits with high dv/dt's should be of shorted emitter construction with its intrinsic high dv/dt withstand capability.

Figure 1-7 shows a simple form of shorted emitter thyristor structure. Externally applied gate current I_G flows from gate to cathode laterally through the gate p-region. The voltage drop developed across the lateral base resistance of p forward biases the right-hand edge of the cathode junction. If gate current is sufficiently large, electrons are injected from this point, and the device turns on in normal pnpn fashion when regeneration begins.

The effect of the partial gate to cathode short is somewhat the same as placing a resistor in parallel with the gate cathode junction of a conventional non-shorted emitter device. This resistor diverts some of the thyristor's thermally generated leakage current and/or dv/dt induced capacitive charging current around the gate-

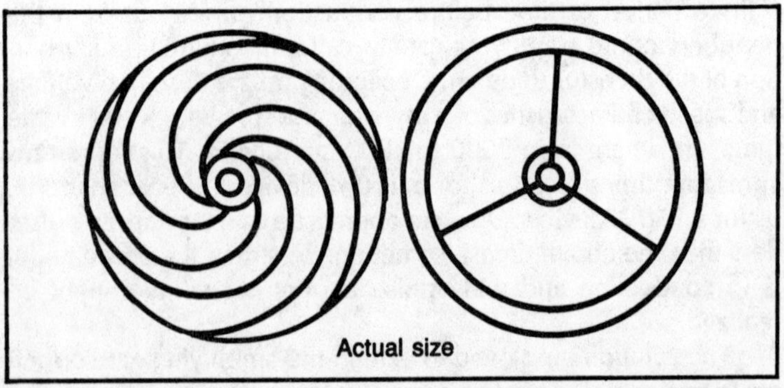

Fig. 1-6. An interdigitated amplifying gate structure (courtesy General Electric).

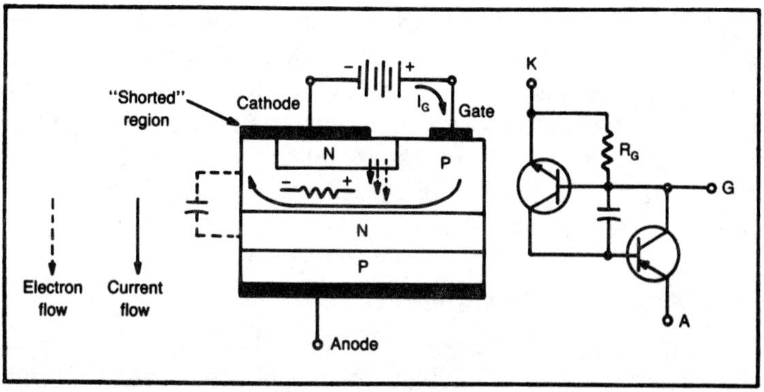

Fig. 1-7. Shorted emitter structure (courtesy General Electric).

cathode junction, by providing an alternative lower impedance path to the cathode. Regeneration is reduced and a shorted emitter thyristor has, as a result, superior high temperature characteristics and dv/dt capability. Emitter shorts reduce the emitter area that can conduct principle current and also interfere with the turn-on of the device, thereby reducing di/dt.

Turn-off Time

It was pointed out earlier that the stored minority charge in the B_n base must decay to zero by recombination before forward voltage may be applied to the SCR without its turning on. This recombination effect can be represented by the simple formula:

$$\left.\frac{dP_n}{dt}\right| \text{recombination} = \frac{-P_n}{\tau_p}$$

where P_n is the excess minority charge (holes in this case) and τ_p is the lifetime of holes in the B_n base. Representative values of τ_p range from 0.1 to 1000 μs and depend upon purity, structure and doping of the silicon. High frequency operation requires short turn-off times so that τ_p must be made small. Usually, this is accomplished by introducing heavy metal or irradiation damage into B_n, which act as additional recombination centers. However, as mentioned before, as τ_p goes down, so do the voltage and current ratings.

Gate Current

In designing thyristors for high frequency applications, almost all of the structure changes that are introduced tend to require greater gate drive. Lowering the carrier lifetime to obtain short turn-off time, introducing a shorted emitter to obtain improved dv/dt, increasing the amount of cathode periphery, using an n^+ gate to reduce delay time—all of these modifications require increased gate current. The amplifying gate mentioned earlier overcomes some of these problems.

High Frequency Operation

High frequency operation stretches all of the SCR ratings to their limits. The circuit subjects the SCR concurrently to high di/dt's, high dv/dt's, and short turn-off times, as well as high voltage and current. From the previous discussion, it should be evident that short turn-off times often conflict with low on-state voltages; shorted emitter construction that allows high dv/dt lowers di/dt. Thus, the device designer's goal is to provide a design with a balanced set of attributes to maximize the power switching capability of the device while maintaining gate trigger control.

TRIACS

A *triac* may be thought of as two silicon-controlled rectifiers connected in a reverse parallel configuration and sharing the same gate circuit. When the device is turned on, both SCR portions conduct, but due to the reverse parallel connection, current may flow through the circuit in either direction. Therefore, the triac may be used for power control of alternating current. This is not possible using the silicon-controlled rectifier.

The basic triac structure is shown in Fig. 1-8. The vertical region directly between terminals MT_1 and MT_2 can be visualized as a pnpn switch in parallel with an npnp switch, or, in other words, as a back-to-back connected pair of reverse blocking triode thyristors (SCRs) with a single external gate connection. In reality, the gate design is quite complex, but it is this complexity that enables the triac to be triggered by a sole control electrode, independent of the polarity of the applied main terminal voltage or even the polarity of the applied gating signal. Figure 1-8 also illustrates a triac symbol oriented in proper relationship to the structural diagram. Since the terms "anode" and "cathode" are not applicable to a triac, main terminal connections are designated simply by number. The ac volt-ampere characteristics of the triac are based on MT_1 as the reference point.

The first quadrant, Q-I, is the region where MT_2 is positive with respect to MT_1 and vice versa for Q-III. The breakover voltage, $V_{(BO)}$, in either quadrant (with no gate signal) must be higher than the peak of the normal ac waveform applied in order to retain control by the gate. A gate current of specified amplitude of either polarity will trigger the triac into conduction in either quadrant, provided the applied voltage is less than $V_{(BO)}$. If $V_{(BO)}$ is exceeded, even transiently, the triac will switch to the conducting state and remain conducting until current drops below the holding current, I_H. This action provides inherent immunity for the triac from excessive transient voltages and generally eliminates the need for auxiliary protective devices. In some applications, the turning on of the triac by a transient could have undesirable or hazardous results on the circuit being controlled, in which case transient suppression is required to prevent turn-on, even though the triac itself is not damaged by transients.

Triac current ratings are based on maximum junction temperature, similar to SCRs. The current rating is determined by conduction drop; i.e., power dissipation and thermal resistance junction to case, and is predicated on proper heat-sinking. If the case temperature is allowed to go above its rated value, as determined from the specification sheet, the triac can no longer be

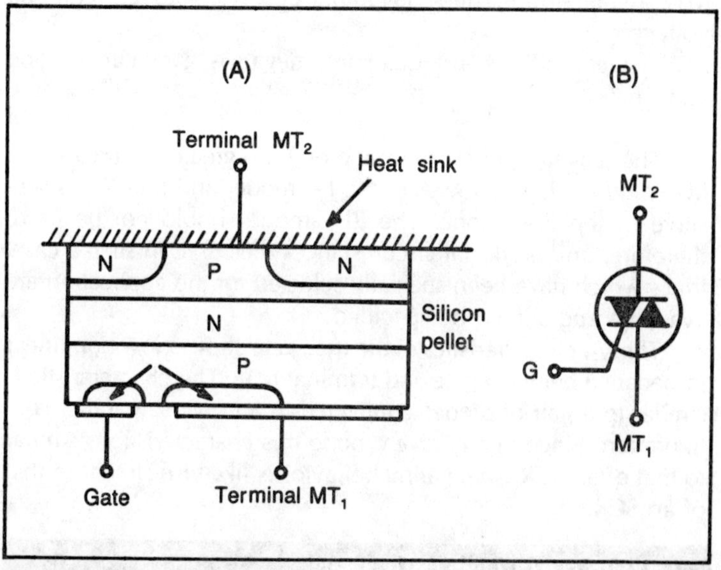

Fig. 1-8. Simplified pellet structure and schematic symbol of a triac (courtesy General Electric).

guaranteed to block its rated voltage, or to reliably turn off when main terminal current goes through zero.

For inductive loads, the phase-shift between line current and line voltage means that at the time that current drops to the I_H value and the triac changes to the non-conducting state, a certain line voltage exists which must then appear across the triac. If this voltage appears too rapidly, the triac will immediately resume conduction. In order to achieve proper commutation with certain inductive loads, the dv/dt must be limited by a series RC circuit in parallel with the triac, or current, voltage, phase-shift, or junction temperature reduced.

Since the triac may be triggered with low energy positive or negative gate current in both the first and third quadrants, the circuit designer has a wide latitude for selecting the control means. Triggering can be obtained from dc, rectified ac, ac, or pulse sources such as unijunction transistors, neon lamps, and switches diodes such as the ST-2 diac, the silicon bilateral switch (SBS), and the asymmetrical trigger switch (ST-4).

The triggering modes for the triac are:

MT2+,Gate+;I+;First quadrant, positive gate current and voltage.
MT2+,Gate−;I−;First quadrant, negative gate current and voltage.
MT2−,Gate+;III+;Third quadrant, positive gate current and voltage.
MT2−,Gate−;III−;Third quadrant, negative gate current and voltage.

The sensitivity of the triac, at present, is greatest in the I+ and III− modes, slightly lower in the I− mode, and much less sensitive in the III+ mode. The III+ mode should not be used, therefore, unless special circumstances dictate it. In such a case, triacs which have been specially selected for the application are available and should be specified.

The V-I characteristics of the triac gate show a low non-linear impedance between gate and terminal MT_1. The characteristic is similar to a pair of diodes connected in an inverse parallel configuration. Since in any given mode this characteristic is similar to that of an SCR gate, gating behavior is likewise similar to that of an SCR.

THYRISTOR CONSTRUCTION

The successful and reliable operation of a thyristor is

predicated on its proper design and construction. The fabrication methods chosen for a particular thyristor type depend a great deal on the service expected from that type. A 70A SCR destined for use in a harsh military environment may differ radically in design from a 6A triac intended for the light industrial or consumer markets.

Pellet Fabrication

The heart of a pnpn device is its multilayered pellet of alternate p- and n-type semiconductor material. This semiconductor is almost always silicon, although germanium has been used. Pellets may be fabricated by any one of several methods, depending on the desired characteristics, complexity, and size of the finished device. The most popular pellet fabrication methods are *diffused* and *planar diffused*. The manufacture of all-diffused pnpn pellets starts with the preparation of large area pnp wafers. These are formed by gaseously diffusing p-type impurities simultaneously into both faces of a thin wafer of n-type silicon. Where specific device characteristics require it, a second diffusion step is used to complete the final pnpn structure. To do this, each pnp wafer is selectively masked (on one side only), and subsequently diffused with n-type impurities through the windows in the mask. The finished pnpn wafers are then diced into individual pellets. Triacs and other more complex structures fabricated using similar techniques. Figure 1-9 shows the various steps leading to the fabrication of a typical diffused SCR structure.

Pellet Encapsulation

Pellet encapsulation methods vary widely, depending on the in-service environment and circuit operating conditions imposed on the finished product, and on the type and size of the chip (or chips) being encapsulated. Two fundamentally different methods are used to house large area pellets. The older, more traditional method is to hard solder (using a gold-eutetic alloy) the anode side of the pellet to a refractory metal (tungsten or molybdenum) back-up plate, the latter matching the thermal expansion of Si. This subassembly, along with a cathode molybdenum back-up plate and two silver shims to reduce thermal resistance, is held firmly in place inside a stud housing by internal Belleville spring washers (see Fig. 1-10). This PIC-PAK construction (Pressurized Internal Contacts) minimizes the possibility of thermal fatigue.

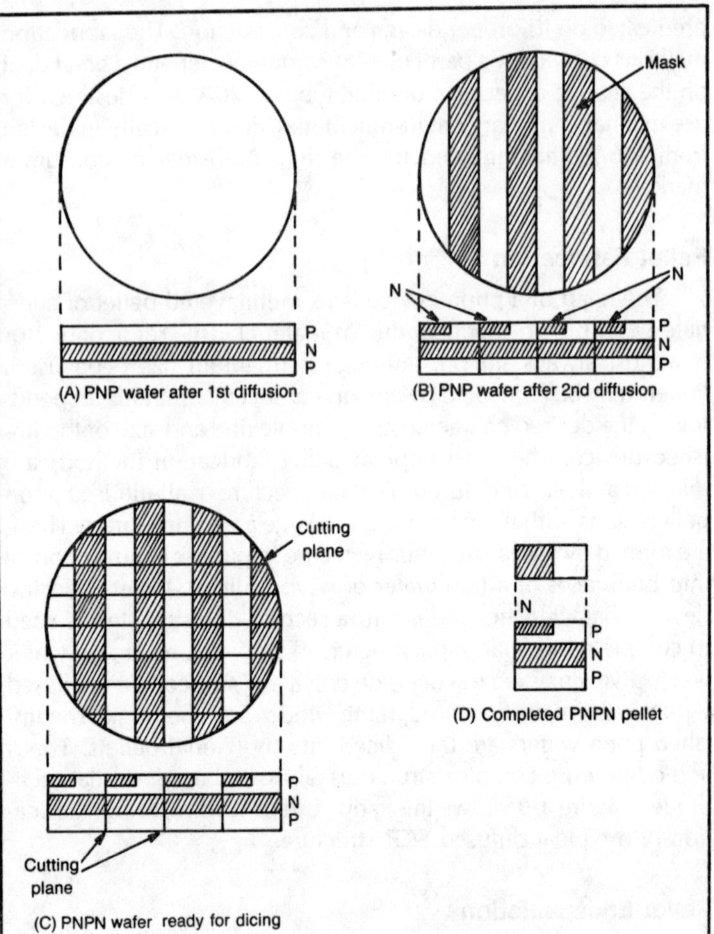

Fig. 1-9. All-diffused pellet formation (courtesy General Electric).

In should be emphasized, however, that while the use of hard solder is virtually indispensable for the (soldered) mountdown of large diameter silicon chips, in smaller devices it is possible to substitute special and lower cost soft solders (usually with a high lead content), without sacrificing thermal cycling capability. Figure 1-11 portrays a power diode that utilizes soft-solder mountdown. Here, for cost and effectiveness, the traditional and expensive ceramic metal housing is replaced by a molded plastic cap that, in conjunction with state-of-the-art pellet passivation (polyimide coatings), provides more than adequate long-term

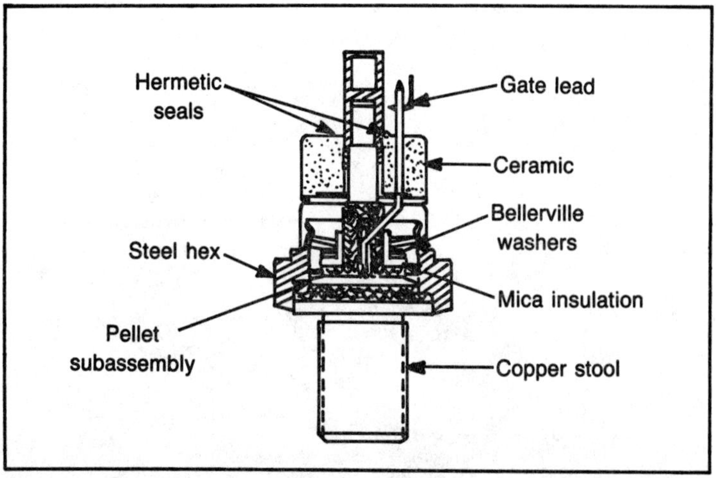

Fig. 1-10. Internal compression bonded construction (courtesy General Electric).

blocking stability. In this GE exclusive HI-LINE design, because the plastic cap is hollow, there is no contact between plastic and silicon. Consequently, both blocking stability and thermal cycling behavior are simultaneously enhanced by elimination of the thermally mismatched plastic-silicon interface that results from classic direct molding, and is extremely susceptible to contamination via capillary action.

The HI-LINE technology extends to far more complex arrangements than the simple diode just described. Figure 1-12 illustrates a HI-LINE Power Module, in which two power semi-

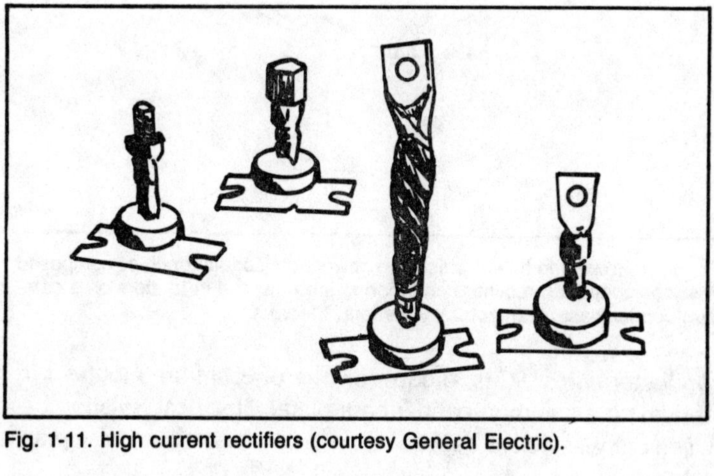

Fig. 1-11. High current rectifiers (courtesy General Electric).

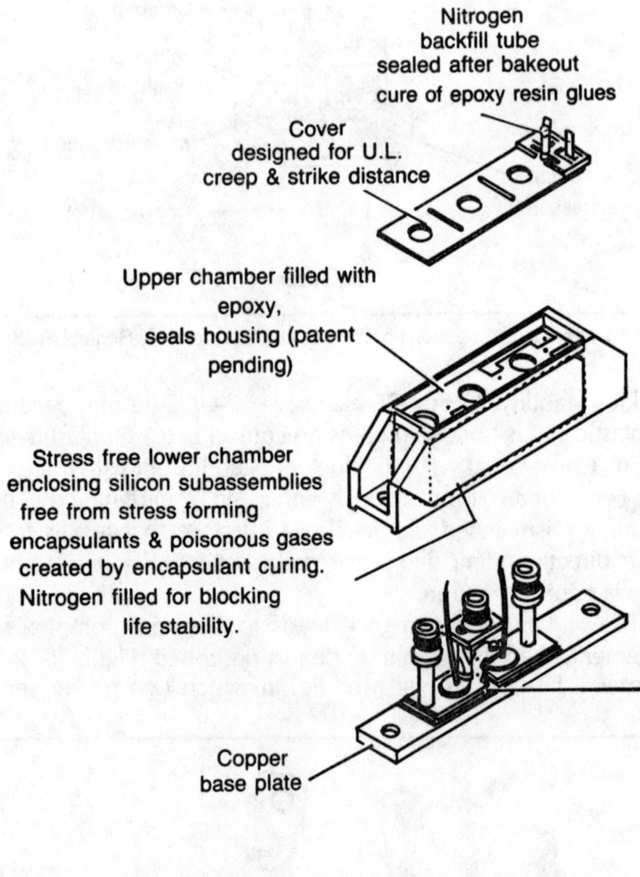

Fig. 1-12. Power module in which two power semiconductor chips composed of silicon-controlled rectifiers and diodes are mounted in tandem on a common copper baseplate (courtesy General Electric).

conductor chips (SCRs, diodes, or one of each) are mounted in tandem on a common copper base plate. Electrical isolation between chips as well as between chips and the common base plate

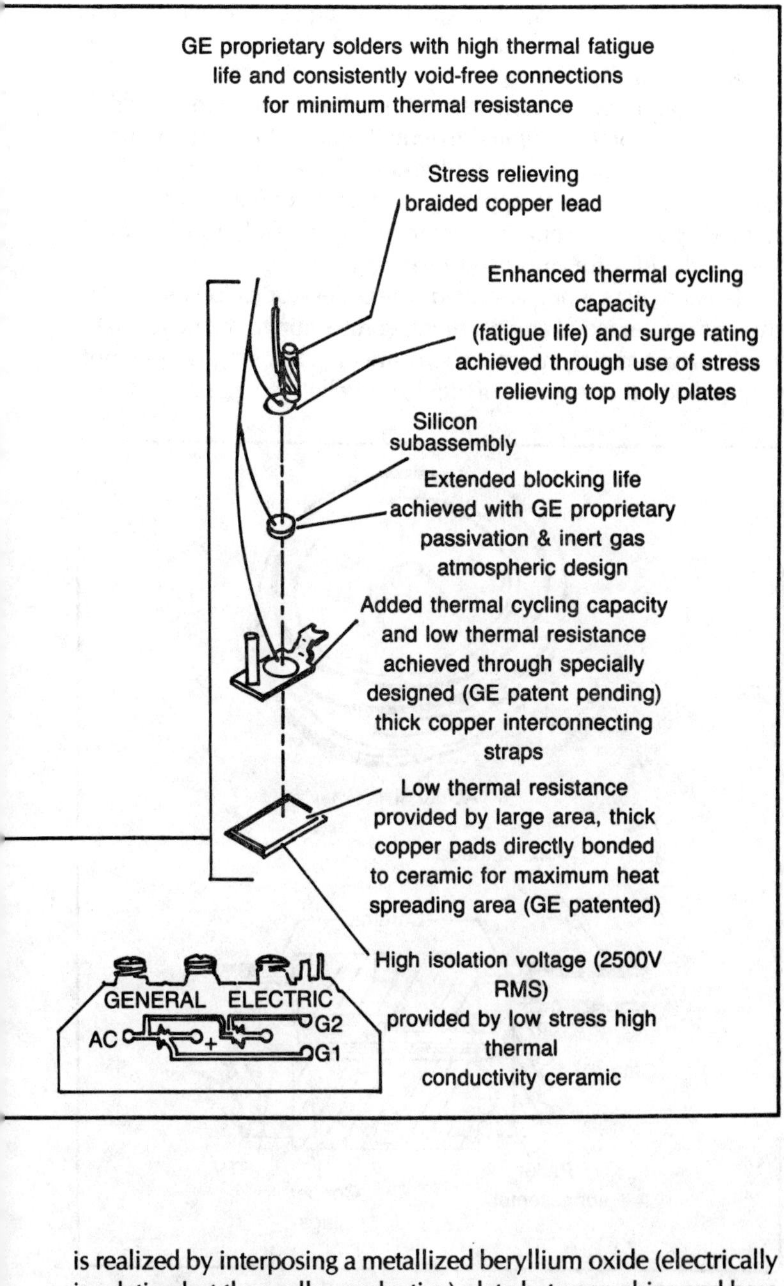

is realized by interposing a metallized beryllium oxide (electrically insulating but thermally conductive) plate between chips and base plate. In this way, because the base plate is electrically isolated, the module may be mounted directly to any convenient cooling medium (an equipment chassis, for example) without incurring cost for additional isolation.

A variation of the compression bonding technique, in which the retaining force is applied externally rather than by internally located spring washers, is the Press Pak, shown in Fig. 1-13. A particular advantage to applying the force externally is that this method allows the semiconductor pellet assembly to be cooled from both sides for improved heat transfer.

Because planar or glassivated type pellet fabrication inherently give a large degree of environmental protection to the pellet electrical characteristics, hermetic glass-metal packaging is often not required for these types of devices. Pellets so passivated are

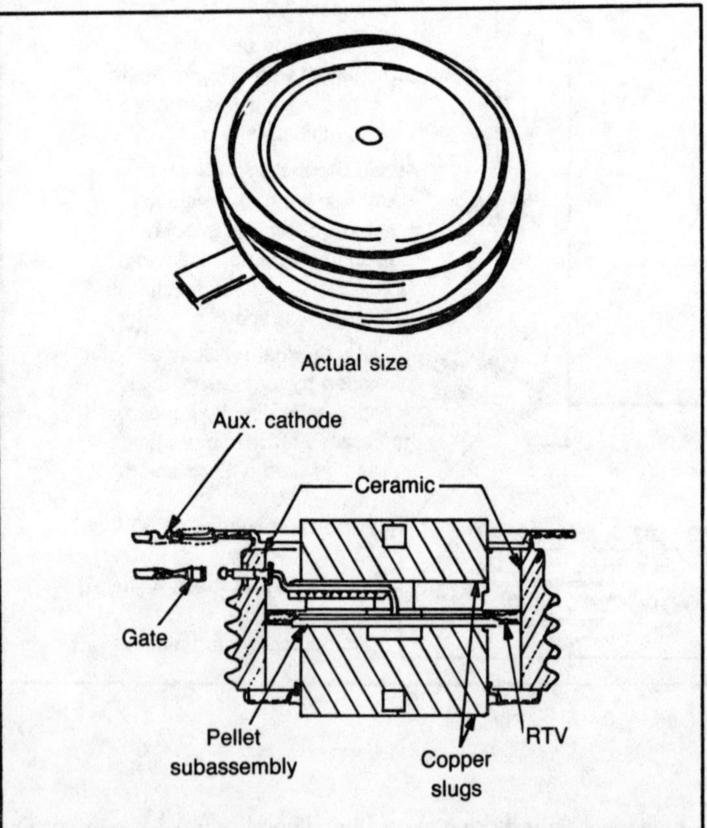

Fig. 1-13. Press Pak encapsulation for the General Electric 550 series of semiconductors (courtesy General Electric).

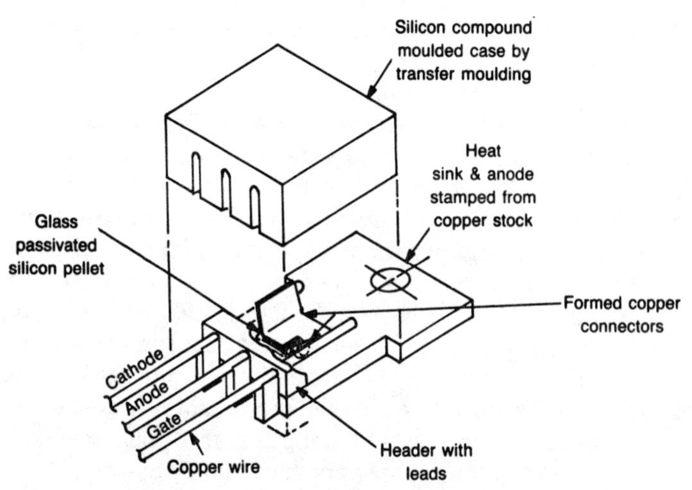

Fig. 1-14. Pictorial view of the Power Pak assembly (courtesy General Electric).

generally encapsulated in a transfer-molded, silicone plastic package, which provides good mechanical integrity to the finished device. Figure 1-14 shows a plastic package used for glassivated triacs and SCRs.

Another long-sought means of pellet passivation has recently become feasible: glass passivation. A thin coating of low temperature glass is fused onto the silicon chip and performs the same function as the SiO_2 layer mentioned above. Glass passivation is superior to SiO_2 (in actuality, also a glass), since it can be applied in thicker coatings, giving superior passivation and allowing higher voltage ratings. Because glassivation is a post-diffusion passivation process, it can be readily applied to mesa-type pellet structures. In general, much higher blocking voltages can be obtained from mesa structures than from the planar process and additionally, silicon utilization (amps/sq cm) is much greater. Thus, mesa structures are more desirable for high voltage, high current devices like large thyristors.

OTHER THYRISTOR DEVICES

It should be apparent at this point that the design of a thyristor, especially one destined for high frequency applications, is necessarily a compromise between various conflicting performance criteria. Fortunately, however, in many fast switching circuits, the thyristor is bypassed in its reverse blocking direction with an antiparallel diode that serves to conduct reactive load currents. In such applications, reverse voltage applied to the SCR is limited to a few volts (see Fig. 1-15).

The principle of the asymmetric thyristor is to capitalize on this circuit characteristic by designing a pnpn structure that dispenses with high reverse blocking voltage and consequently offers significantly improved performance as far as other parameters are concerned. The asymmetric structure differs from that of a conventional SCR, as shown in Fig. 1-16, by the interposition of the highly doped N+ layer between the P+ anode and the normal N base. The purpose of this layer is to prevent spreading of the electric field that exists in the N base during forward blocking into the anode P+ region where it would precipitate voltage-induced forward breakover. In conventional SCRs, this phenomenon is avoided by making the N-base very thick (more spreading room) with consequent deleterious effects on forward voltage drop. In the ASCR, the N-region is thin, and forward drop is considerably lower. Should the prospective user feel that minimum on-state losses are not essential, the forward drop can be brought back to normal SCR levels by additional gold doping or electron irradiation, which has the merit of causing a dramatic decrease in turn-off time. An intermediate course is to adjust doping levels for (lesser) gains in both parameters. Since reverse blocking of the asymmetric structure is dependent on the P+N+ junction, which is bounded by zones of low resistivity, the overall reverse blocking capability is limited to 10-50 volts.

There is a whole family of light-activated thyristors. These operate in much the same manner as the components discussed

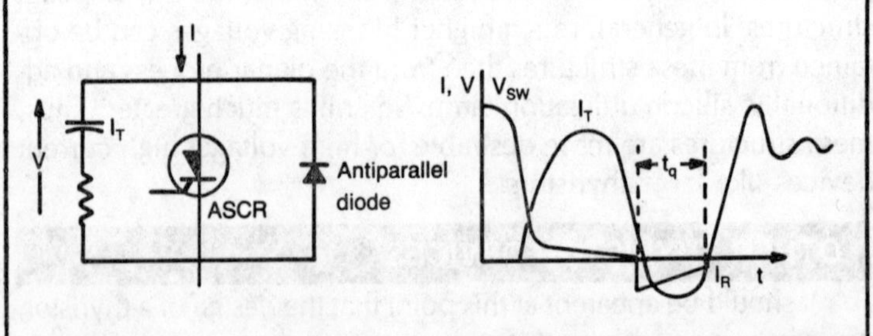

Fig. 1-15. Waveforms for sinusoidal currents flowing through an asymmetrical silicon-controlled rectifier (courtesy General Electric).

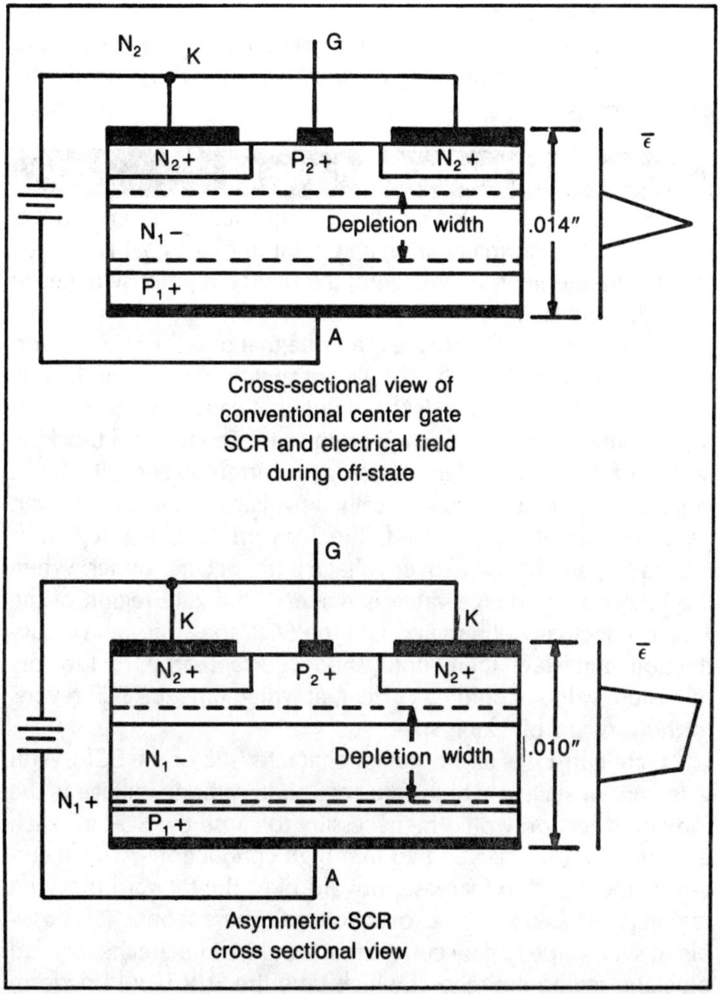

Fig. 1-16. Characteristics of reverse blocking triode or tetrode thyristors (courtesy General Electric).

previously. However, a light source is used as a means of causing these components to conduct current. The LAS is a light-activated switch, or light-activated reverse blocking diode thyristor. The LASCR is a light-activated silicon-controlled rectifier whose gate circuit is set up in the conventional manner, but in a controlled way which lets various intensities of light cause the device to break into forward biased conduction. Similarly, there is the light-activated silicon-controlled switch, which contains four electrodes, and even light-activated triacs.

All of these components and many others are derivatives of the original silicon-controlled rectifier. All fall in the general category of thyristors.

SILICON-CONTROLLED RECTIFIERS

The silicon-controlled rectifier is a three-junction semiconductor device. It is normally an open circuit until a signal is applied to the gate element. At this time, the device rapidly switches to a conducting state.

Forward blocking voltage is a value that puts the SCR in a forward biased state, but which does not switch it to the conducting state. Breakdown voltage is that value that causes the SCR to actually switch to the conducting or on state. The forward blocking region of the SCR is largest when gate current is zero. It takes a greater value of forward anode/cathode voltage to cause breakover. As gate current is increased, the forward blocking region is decreased, and the breakover voltage point occurs sooner. When the breakdown voltage value is reached, the gate region of the SCR is effectively eliminated and the SCR goes into heavy conduction and stays there until anode current drops below the minimum value of holding current, at which time the SCR reverts to the forward blocking state.

Gate current is an important characteristic of the SCR. With gate current at zero, a high value of anode/cathode voltage in the forward direction would be necessary to cause the SCR to reach the breakover voltage and go into high conduction. As gate current is increased to increase forward bias, the forward blocking region is reduced and breakover voltage occurs sooner. It is possible to set a value of gate current that would eliminate the forward blocking region entirely, at which time the SCR would perform as a conventional PN rectifier.

Once the SCR reaches the high conduction region, the gate has no control and the gate has been effectively eliminated. To regain control, external methods must be applied to the SCR to reduce the flow of anode current. One method is to apply a voltage to the anode that will make the anode negative with respect to the cathode. This turns the SCR off, putting it back to a forward blocking condition. Another method is to divert enough anode current to some external circuit for a period of time long enough to cause holding current to drop below its minimum value, thereby turning the SCR off. A switch across the anode/cathode terminals, electronically activated or otherwise, will divert the anode current flow. Still another method of turning the SCR off is to cause a heavy reverse current flow from gate to cathode by injecting a reverse voltage which will cause the gate region to increase its opposition to anode current flow and allow the gate to regain control of the device.

The blocking voltage capability of all semiconductor junctions is temperature sensitive. The avalanche voltage (the ultimate voltage limit) decreases with temperature. This necessitates a voltage derating for measurements taken at room or elevated temperature. A second factor that determines a temperature limit is to limit thermally induced mechanical stresses in the silicon crystal to safe values.

The maximum junction temperature limit is imposed because of the positive temperature coefficient dependence of intrinsic leakage currents and increased turn-off time. Intrinsic leakage currents increase exponentially with temperature. For a simple, reverse biased pn junction, increased leakage currents result in higher power dissipation, which could lead to thermal instability. In thyristor structures, high leakage current decrease off-state voltage capability and increase turn-off times.

The rated maximum operating junction temperature limit is used to determine steady-state and recurrent overload capability for a given heatsink system and maximum ambient temperature. Conversely, the required heatsink system may be determined for a given loading of the semiconductor device by means of the classic thermal impedance approach.

Transiently, the device may actually operate beyond its specified maximum operating junction temperature and still be applied within its ratings. An example of this type of operation occurs within the specified forward non-recurrent surge current rating. Another example is the local temperature rise of the junction due to the switching dissipation during the turn-on of a thyristor under some conditions. It is presently impractical to establish temperature limits for these types of operating stresses from both a rating as well as an applications point of view. Therefore, such higher-than-rated temperature operation must remain implicit in other ratings established for the device.

The power generated in the junction region in typical device operation consists of the following five components of dissipation:

- Turn-on switching.
- Conduction.
- Turn-off or reverse recovery.
- Blocking.
- Triggering (for thyristors).

The silicon-controlled rectifier is a most remarkable solid-state device, and while modifications have been made, most designs are basically identical to the first component made more than 25 years ago. Probably the most pleasing aspect of the SCR is its simplicity, especially in regard to its use in electronic circuits. Many, many components (electronic and mechanical) can be eliminated from all types of equipment through the use of the silicon-controlled rectifier.

Chapter 2

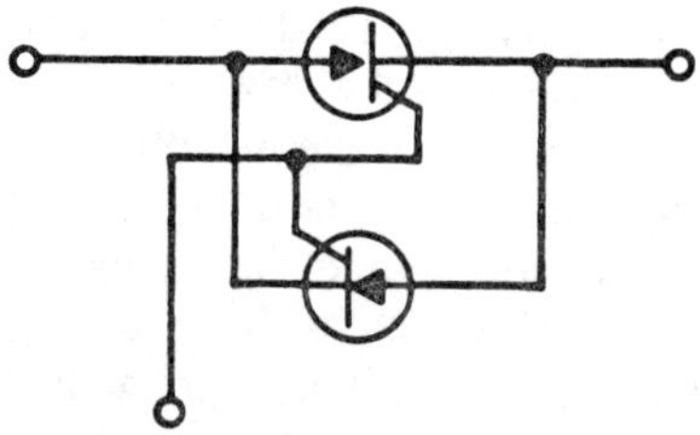

Electronic Components

WHILE THIS IS A BOOK ON SCR PROJECTS, YOU MUST REMEMBER that other electronic and electrical components are needed to complete most SCR circuits. All of the projects involve silicon-controlled rectifiers, but may also use transistors, pn junction diodes, transformers, resistors, capacitors, inductors, and other such common devices.

In order to build the projects in this book, you must have a basic understanding of not only silicon-controlled rectifiers, but also the other components required for circuit completion. You will also have to know how to test these devices in order to troubleshoot any project which does not seem to be operating properly.

This chapter provides you with a basic understanding of electronic components and circuit values. This is not an in-depth, theoretic, technical study of how electronic components work, but rather an overview of what each common component does. Each component will be discussed in simple terms, with special emphasis on how it operates within a basic circuit. If you are new to electronics and project building, this chapter will give you the groundwork of components and circuit assembly that you'll need. If you are an experienced builder, you will probably find the information presented here to be a helpful refresher course.

VOLTAGE

Voltage can be described as potential difference. Voltage does not flow; rather, it is a force that causes a flow of current in a circuit. It can be compared to a bat that strikes a baseball. The movement of the ball is similar to the flow of current, and the bat is the potential force that causes the movement of the ball.

Voltage is either alternating or direct. Direct current is most often provided by batteries, while alternating current is obtained from standard receptacles such as those found in homes.

While the volt is a basic unit of potential difference, there are a number of derivatives of the volt which are used to keep the numbers more practical. The term *millivolt* is used to describe potential differences which are equivalent to 1/1000th of a volt. An even smaller quantity is the *microvolt*, which is 1/1000th of a millivolt, or a millionth of a volt. One *kilovolt* is equal to 1,000 volts. Most of the projects in this book involve potential differences which can be measured in volts.

When speaking of voltage, it is necessary to know the type of voltage under discussion. In basic electrical circuits, voltage is measured between the positive and the negative pole. Contacts from both of these poles must be applied to the circuit to be powered or to the measuring device in order to establish a flow of current. When properly connected to an electronic circuit, or load, a power source will cause current to flow from the negative to the positive pole.

All direct current *(dc)* voltage circuits have *filed poles*. This means that the polarity remains constant. In other words, one contact is always positive and the other is always negative. In contrast, the polarity of alternating current is continually being reversed. During one half of the ac cycle, one pole is positive while the other is negative. During the next half, the pole that was positive switches to negative, and the negative pole becomes positive. The rate of polarity change in ac circuits,or *alternation* is measured in hertz, which is equivalent to the number of cycles per second. Household power is alternating current and reverses at a rate of 60 hertz. This means that the polarity reverses itself 60 times each second. This rate of change is called the *frequency*.

CURRENT

Current flow is measured in *amperes*. The term "flow" is quite accurate when discussing current, as electrons are caused to flow through a circuit when a voltage source is connected. Another term which is quite accurate is "circuit," because to establish a flow of current, it is necessary to have a complete electrical path between the positive and negative poles of the voltage source. When this path, or circuit, is interrupted, the current flow ceases.

An ampere is actually an indication of the number of electrons which move past a specific point in a circuit in a given period of time. The ampere is equivalent to approximately six trillion electrons per second. In practical measurements, the number of electrons is not stipulated, but their movement is expressed in amperes or in fractional units thereof. The flow of electrons is always from the negative to the positive pole.

As with the volt, there are many derivatives of the ampere in order to express it in practical terms. *Milliampere* is often used to describe the current values found in solid-state electronic circuits and is equal to 1/1000th of an ampere. The *microampere* is equal to 1/1,000,000th of an ampere, or 1/1000th of a milliampere. This term is used to describe the current values of extremely low-powered electronic circuits and signals.

The measurement of current is accomplished by placing an ammeter or milliammeter directly within the current path of a circuit. This is unlike measuring voltage, which is done by placing a multimeter probe at the positive and negative poles of the voltage source.

CONDUCTORS

A *conductor* is any material which provides a path for the flow of current. Common electrical wire is probably the best example of a conductor, as it allows electrons to pass freely through its copper material. Frequency has a great deal to do with the conductivity of a material. A material that easily conducts the flow of direct current or low-frequency alternating current will often begin to resist the flow of current as the frequency increases. Copper wiring, then, can be thought of as a true conductor.

There is no such thing, however, as a perfect conductor. Any material that conducts electrical current also resists this same flow to some small degree. When speaking of copper wiring, the resistance to current flow increases as the electrical path is lengthened. For example, a small battery easily passes enough current through a miniature light bulb to cause it to glow. If the battery is connected to the bulb through a copper wire that is several thousand feet long, the resistance is probably great enough to prevent an adequate amount of current to pass through the bulb's filament, and the bulb would not get enough current to light up.

RESISTANCE

The term resistance describes the opposition of a material to the flow of current and is measured in ohms. Electronic circuits used calibrated, lumped resistances to control the flow of current. These are called resistors and come in many different configurations. Figure 2-1 shows some of the more common types of resistors found in electronic circuits. These are usually quite small devices which provide a specific resistance to the flow of current in a circuit. Most of the fixed resistors used for the projects in this book will be carbon devices. Carbon resistors are available in half-watt, one-watt and two-watt power ratings. The physical size of the resistor determines its capability to handle power but not its resistance. A physically large resistor may have a very high power rating

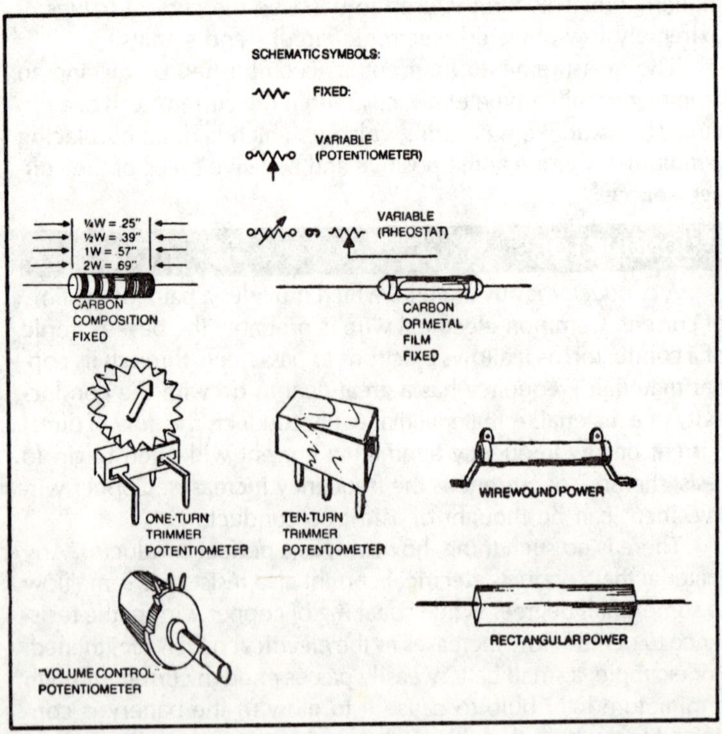

Fig. 2-1. Schematic symbols and pictorial drawings of several types of common resistors.

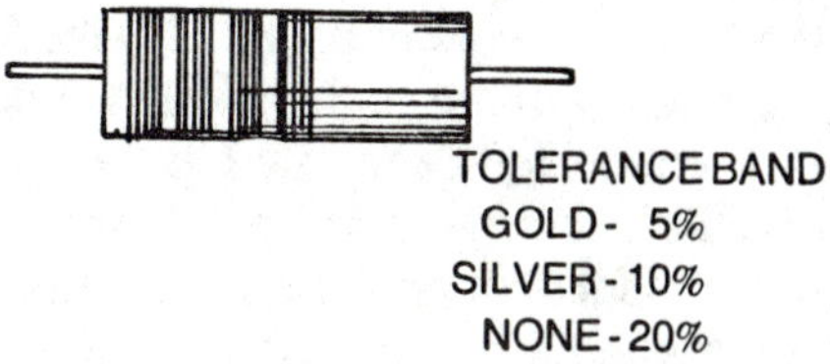

COLOR	FIRST BAND	SECOND BAND	THIRD BAND (MULTIPLIER)
BLACK	0	0	1
BROWN	1	1	10
RED	2	2	100
ORANGE	3	3	1000
YELLOW	4	4	10,000
GREEN	5	5	100,000
BLUE	6	6	1,000,000
VIOLET	7	7	10,000,000
GRAY	8	8	100,000,000
WHITE	9	9	

Fig. 2-2. The resistor color code chart.

and a very low resistance, whereas a small resistor may have a resistance of several million ohms and a very low power rating.

The resistance value of a carbon resistor is most often indicated by colored bands. (Sometimes, the value is indicated by a number stamped on the side of the device, but this is quite rare.) It is important for the experimenter to know how to read the color coding designations on resistors in order to determine their values. Figure 2-2 shows a standard color code chart that will allow you to read the values of any conventionally coded carbon resistors.

POWER

Power is most often defined as the rate of doing work. In electronic applications, power is measured in *watts*. In electrical circuits, power is equal to the potential (voltage) across the component or circuit multiplied by the current that flows through it. The formula which explains this is $P = EI$, where P equals power in watts, E equals potential in volts, and I equals current in amperes. For example, if an electrical potential of 9 volts is applied to a component or complete circuit which causes 0.5 ampere of current to flow, the total power would be 9×0.5, or 4.5 watts.

In any pure resistance, including that of a wire conductor or an actual resistor, the work or power is in the form of heat which first raises the temperature of the resistance device and then, in some part, is conducted or radiated to the air and other nearby objects. Because excessive temperatures are damaging, the rating of a resistor includes, in addition to its resistance, its ability to radiate power without becoming too hot. This is stated in watts. You will find resistors rated for 1/4 watt, 1/2 watt, 1 watt, 2 watts, 5 watts, 10 watts, 25 watts, and higher.

Power in watts may be expressed by the square of the current flowing through a component times its resistance, or $P = I^2R$, where R equals resistance in ohms. Thus, a 10-ohm resistor which is called upon to pass a current of 2 amperes must have a power rating of at least $2^2 \times 10$, or 4×10, which is equal to 40 watts. Also, notice that for the resistor to pass that current, the applied voltage must be 20 volts (since $E = I \times R$). Power may also be calculated by the formulas $P = \frac{V^2}{R}$ or $P = VA$ (voltage times current).

CAPACITANCE

To understand the concept of capacitance, consider a pair of metal plates which are placed near each other, but not touching. If a battery is connected to these plates (the positive pole to one and the negative pole to the other), electrons from the battery will flow to the plate connected to the negative terminal. At the same time, excess electrons will be attracted from the plate connected to the positive terminal of the battery. If the battery is disconnected, one plate will be left with an excess of electrons and the other with a shortage of electrons. Thus, a potential or voltage difference will exist between the two. These plates act as a capacitor. Capacitance is the storage of electrical energy by objects insulated from each other. The amount of energy, or the total number of electrons that can be stored in a capacitor, is proportional to the original voltage, the area of the plates, the distance between them, and the insulating material (dielectric) between the plates.

Capacitance is measured in *farads*. A farad is a very large quantity which is seldom found in electronic circuits, so fractional units are again necessary. A *microfarad* (μF) is a millionth of a farad, and a picofarad (pF) is a millionth of a microfarad. Figure 2-3 shows some of the more common types of capacitors found in electronic

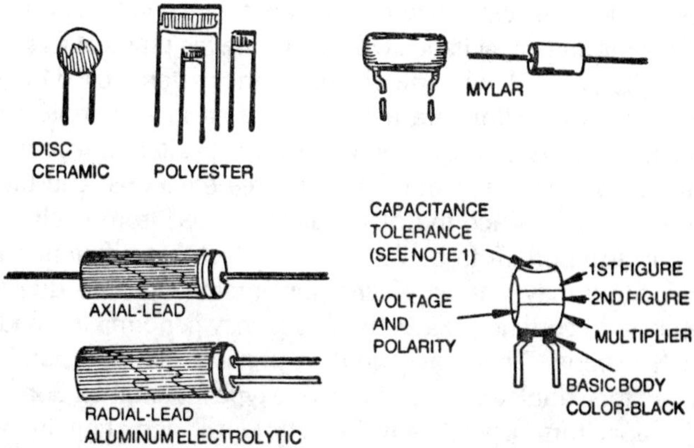

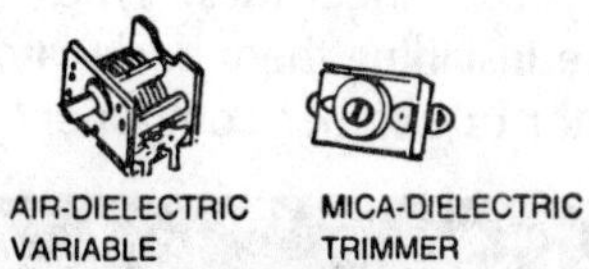

SCHEMATIC SYMBOLS:

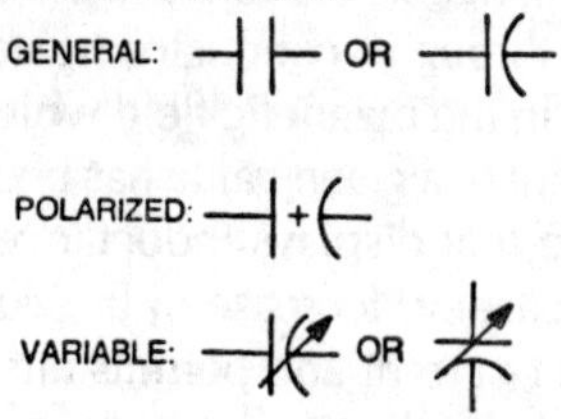

CAPACITANCE AND VOLTAGE COLOR CODE

COLOR	+85 C Voltage	Cap. in Picofarads		Multiplier
		1st Fig.	2nd Fig.	
Black	4	0	0	—
Brown	6	1	1	—
Red	10	2	2	—
Orange	15	3	3	—
Yellow	20	4	4	x10^4
Green	25	5	5	x10^5
Blue	35	6	6	x10^6
Violet	50	7	7	x10^7
Gray	—	8	8	—
White'	—	9	9	—

SOLID TANTALUM ELECTROLYTIC (SOME MAY BE MARKED WITH PRINTED NUMBERS RATHER THAN COLOR CODE)

Fig. 2-3. Schematic symbols for and pictorial drawings of several types of capacitors.

projects, as well as their corresponding schematic symbols.

Capacitors are rated in terms of operating voltage, capacitance, and sometimes temperature coefficient. The combination of voltage rating and capacitance will determine the physical size of each capacitor, as will the particular dielectric material used in its construction. Capacitors are used quite often in electronic circuits, because they have the capability of passing alternating current (as the plates alternately gain and lose free electrons) and blocking direct current (since the plates are insulated from each other).

It is important to be aware of the fact that different types of capacitors may not be electrically interchangeable due to differences in current leakage, high frequency performance, and other such criteria. Therefore, until you gain an understanding of capacitors, don't try to substitute one type for another. Some types of capacitors (specifically, electrolytics and tantalums) are polarized, and their cases will either be marked with a dot or plus (+) sign (positive terminal) or a negative or minus (−) sign (negative terminal). It is very important that you connect these types of devices in the proper direction, since installing them backwards can easily destroy either the capacitor or another component.

INDUCTANCE

Inductance is the opposition to change in the rate of flow of an electric current. The flow of current through a conductor creates a magnetic field, and energy is stored in the magnetic field which surrounds the conductor. Once a current of a given value has been established in an inductor (the device that displays inductance), that inductance will oppose any increase or decrease in the current. Just as a capacitor blocks a direct current and permits alternating current to pass, an inductor allows direct current to pass while opposing the flow of alternating current, because the alternating current changes faster than the magnetic field can change.

The henry is the unit of inductance. For many applications, the term millihenry is used, which is abbreviated mH and is equal to 1/1000th of a henry. Also common is the microhenry (μH), which is a millionth of a henry.

Inductors can be made in many different ways, depending upon the amount of inductance required in a circuit and the frequency at which the inductor is to operate. For radio frequencies, an inductor can be a simple coil of a few turns of wire or wire which has been wound around a ferrite (powdered iron) core. At audio and power line frequencies, inductors generally resemble

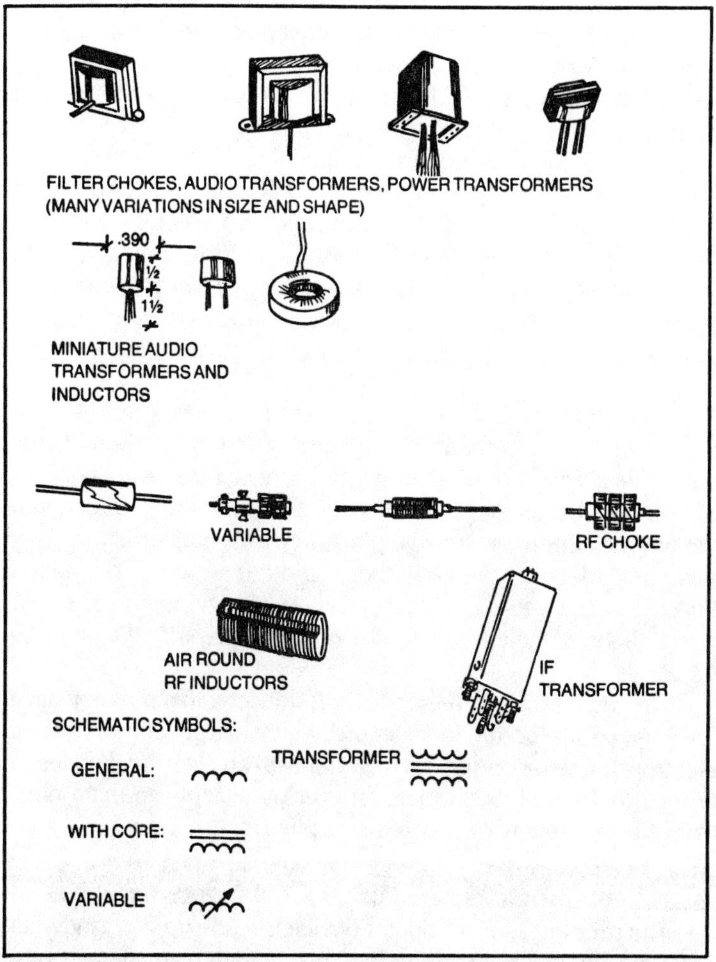

Fig. 2-4. Transformers and inductors.

audio and power transformers; that is, many turns of wire which have been wound around a laminated steel core. Figure 2-4 shows some types of inductors and the schematic symbol for an inductor.

SEMICONDUCTOR DEVICES

In conductors like copper, electrical current is carried by electrons that are almost perfectly free to move unimpeded through the atomic structure of the material. In insulators such as glass, ceramic, or the materials used for substrates for circuit boards, the electrons are tightly bound in the atomic structure, and no current can flow unless this structure is disrupted by very high voltages. But there is an intermediate class of materials called *semiconductors*. In these devices, electrons are normally not available to carry current, but they are held in the atomic structure loosely and can be moved if a moderate voltage is applied.

Silicon is the most common material used to make semiconductor devices today. Pure silicon, the major constituent of beach sand, is a good insulator. An atom of silicon has four electrons in its outermost orbit, or shell, and pairs of these electrons are shared with the outermost orbits of neighboring silicon atoms. This means that each silicon atom is actually surrounded by eight electrons which are tightly bound in the atomic structure of the silicon crystal.

In order to use silicon to make semiconductors, carefully controlled amounts of impurity atoms are added to its crystal structure. This is called doping. If the impurity is, for example, atoms of phosphorus, an element with five electrons in its outer shell, the extra electron has no real place in the crystal structure. This additional electron can easily be forced out of place and moved through the structure by the application of a few tenths of a volt. Such silicon is called *N-type* (for negative) *semiconducting material*.

On the other hand, if the silicon is doped with atoms that have fewer electrons (boron, for example, which has only three valence electrons), a semiconducting material with an electron deficiency (P-type) is formed. Such material quickly accepts an extra electron if a few tenths of a volt are applied.

DIODES

The simplest semiconductor device is made up of a sandwich of P-type and N-type semiconducting material, with contacts provided to connect these layers to an external circuit. This is known as a *junction diode*. If the positive terminal of a battery is connected to the P-type material (cathode), and the negative terminal is connected to the N-type material (anode), the loose electrons in the N material are easily forced into the spaces in the P material, and a large current flows. This is called the *forward current* and is almost as great as the current that would flow in a conductor such as copper. If not controlled, this current can become high enough to destroy the diode. When connected with the positive terminal of the battery to the cathode and the negative terminal to the anode, the diode is said to be *forward biased*.

If the connections are reversed, with the positive terminal of the battery connected to the anode and the negative terminal to the cathode, very little current flows because the P-type material accepts the electrons from the negative terminal of the battery, and the N-type material gives up its free electrons to the battery. This results in a state of electrical equilibrium, since the N-type material has no more electrons to give and the P-type material has no more room for electrons. Actually, because the P-type material has a few free electrons and the N-type material has a few spaces or holes, there is a small current when the diode is reverse biased. This is known as the *leakage* or *reverse current*.

The basic characteristic of a semiconductor diode is that it acts like a very low resistance when it is forward biased and a very high resistance when reverse biased. If an alternating current is applied to this device, current flows freely during one half cycle (when the diode is forward biased) and not at all during the other half cycle (when the diode is reverse biased). This makes the semiconductor diode an effective rectifier which can convert alternating current into pulsating direct current.

Although this discussion has dealt only with the junction diode, a diode action can also be achieved by placing a very fine wire (catwhisker) in contact with a slab of semiconducting material of either N- or P-type. This creates a point-contact diode, which is commonly used in radio frequency applications.

There are also many other types of diodes. Some are used for rectification in power supplies, while others are used to rectify and detect small signals. Generally, diodes are specified in terms of forward voltage (the maximum voltage which can be applied in the forward condition), reverse voltage (the maximum amount of voltage that can be applied to the diode in the reverse condition, or how much voltage the diode can safely block), and safe current-handling capability.

Figure 2-5 shows the basic construction of the junction and point-contact diode, schematic symbols, and some types of diode packages. Note that the cathode end of most diode packages is marked with a color band, a dot, or by a series of color bands that identify the type number.

TRANSISTORS

If a third layer of semiconductor material is added to a junction diode, thus making it two N-type layers separated by a P-type layer, the resulting device is known as a *transistor*; in this case, a *bipolar transistor,* because its operation depends upon the flow of electrons in one direction and holes or empty spaces in the other direction. A bipolar transistor is really two junction diodes formed back to back.

Each of the three layers of a bipolar transistor has its own contact for connection to an external circuit. The *emitter* contact is connected to one of the N-type layers and emits electrons when the transistor is operating. The contact to the other N-type layer is called the *collector,* since it will be collecting electrons, and

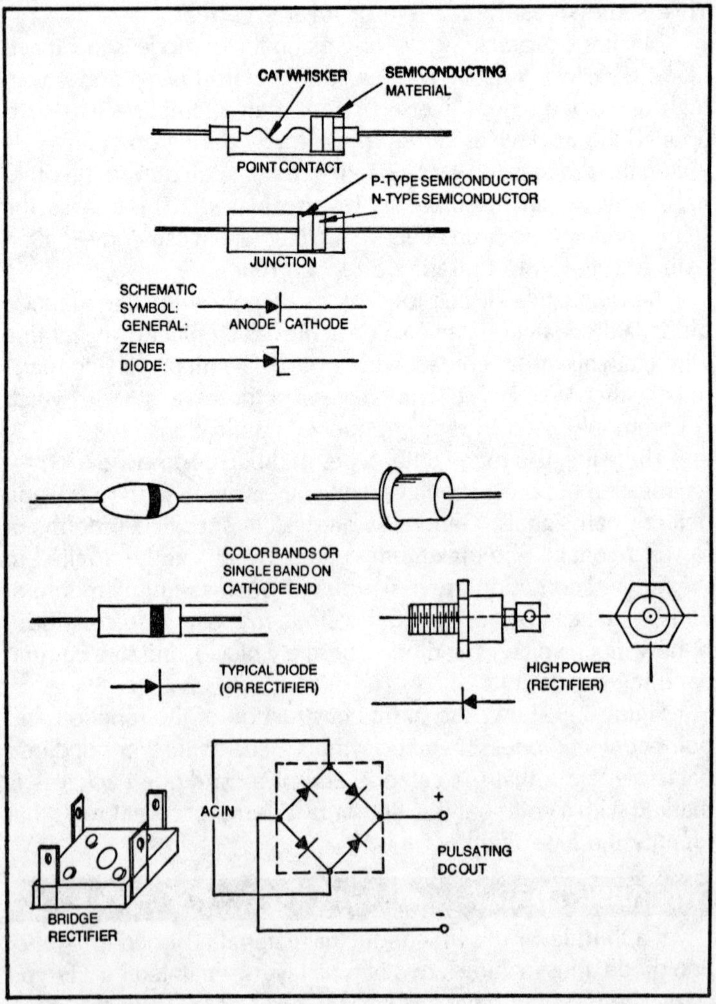

Fig. 2-5. Common diode packages, construction and schematic symbols.

the contact to the intermediate P-type layer is known as the *base*.

In operation of the NPN transistor (which has two N-type layers separated by a P-type) described here, a positive voltage is applied to the collector, a lower positive voltage to the base (about one-half the collector voltage in many instances), and the negative terminals of both voltage sources are connected to the emitter. Here, the emitter and base are connected as a forward biased diode, so you can expect a large current to flow in the emitter-base circuit. The collector and base are connected as a reverse-biased diode, so you can expect little or no current to flow in that circuit. However, some of the electrons that have moved into the intermediate P-type material of the base actually move toward the collector and flow to the higher positive voltage source, rather than to the lower voltage connected to the base.

Because the collector is reverse biased, the collector-to-base resistance is high, but in operation, the current flowing into the emitter and out of the collector are virtually the same, so the power in the circuit is actually higher than that in the emitter circuit. This is evidenced by the relationship shown in the formula $P = I^2R$. Thus, we actually have a gain in power, and this is known as *amplification*.

If the positive voltage applied to the base is reduced, fewer electrons will flow from the emitter to the collector, because there are fewer empty spaces to attract the electrons. If the base voltage is reduced to zero or made negative with respect to the emitter, no current will flow (except for a very slight leakage current), as in the reverse-biased diode. Thus, a small current change in the emitter-to-base circuit is actually controlling a far higher current in the emitter-to-collector current.

If the sandwich is reversed, we have produced a PNP transistor. This device operates in exactly the same way, except that the collector and base are biased with negative voltages with respect to the emitter.

Figure 2-6 shows some of the more common types of transistor packages. Note that although the emitter, base, and collector leads are indicated in these drawings, different lead combinations are possible. Always check the connections on a transistor by referring to a replacement catalog or data sheet before substituting one type for another. Also shown in this figure are the basic structure of PNP and NPN transistors, as well as their respective schematic symbols. Transistor performance is specified in gain, frequency response, power dissipation, maximum voltages, and

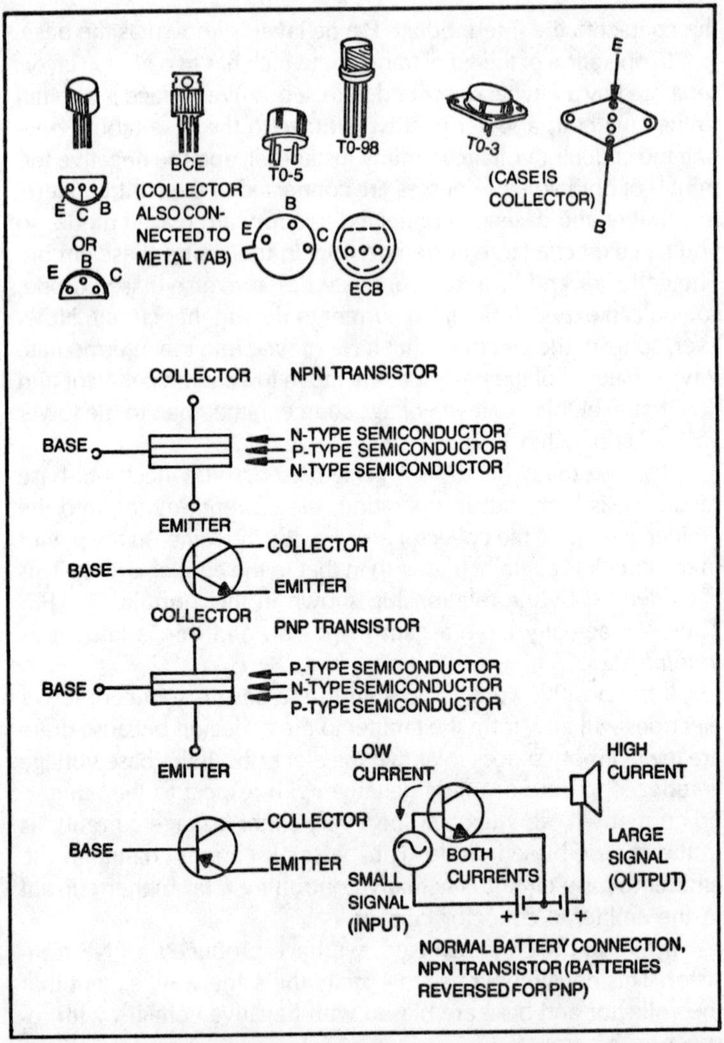

Fig. 2-6. Various types of transistors, their schematic symbols and how they are constructed.

maximum currents. In general, the larger the transistor package, the more power it can safely control.

SILICON-CONTROLLED RECTIFIERS AND TRIACS

The silicon-controlled rectifier (SCR) upon which this book is based is quite similar to the junction transistor in some ways. It is made up of a sandwich of four semiconductor layers in the form of PNPN or NPNP, and has three connections: a cathode and an anode (as with the diode), and a gate. If an SCR is connected in a circuit in the same manner as a conventional diode (positive terminal of a battery connected to the cathode and negative terminal to the anode), it will not normally conduct current until a small positive voltage is applied to the gate. Once the SCR begins to conduct in the forward direction, the gate has no more effect, and the SCR will continue to conduct until the current is removed; that is, until the battery between the anode and cathode is disconnected or until the polarity of the battery is reversed (negative to cathode; positive to anode). This means that if an SCR is connected in an ac circuit, it will conduct only every half-cycle like a conventional diode.

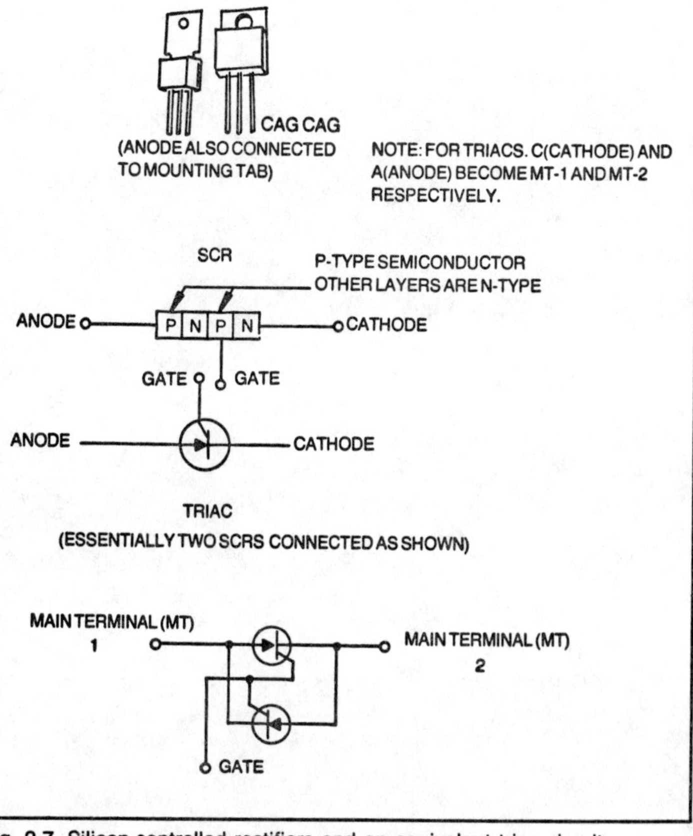

Fig. 2-7. Silicon-controlled rectifiers and an equivalent triac circuit.

SCRs are rated in terms of voltage, current and power capabilities. They are used in a wide variety of circuits as a dc or ac switch, controlled by a small voltage in the gate circuit. A variation called the *triac* is basically equivalent to two SCRs connected back to back. When used to control alternating current, a triac conducts during both halves of the alternating cycle. Figure 2-7 shows the basic construction of the SCR, some typical packages, and the schematic symbols for the SCR and the triac.

Chapter 3

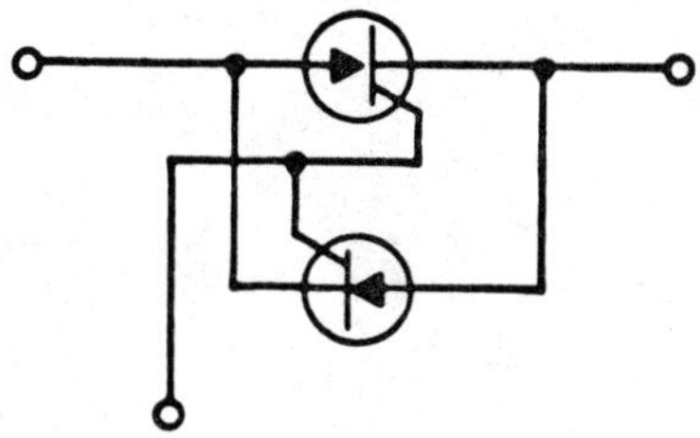

Electronic Measurement

EACH ELECTRONIC PROJECT IN THIS BOOK IS EDUCATIONAL, PRACTIcal, fun to build, and quite inexpensive. Many simple projects found elsewhere may require expensive test instruments for alignment and/or troubleshooting, but this is not true with these projects. Each can be tested and adjusted with a few simple electronic test instruments. In almost every instance, all that is required is a *multimeter,* also called a *volt-ohmmeter,* which measures ac and dc voltage, current, and resistance in ohms.

Any multimeter offers a wide range of measurement capabilities for all types of electronic circuits. This instrument is used by technicians, engineers, and hobbyists alike. It is essential that the beginning experimenter have a good working knowledge of this instrument, as well as an understanding of all that it is capable of doing.

Although there are many makes and models of multimeters, the most basic type is known as the *analog volt-ohm-milliammeter* (VOM), which is equipped with a needle that sweeps across a calibrated scale. One of the more popular VOMs is shown in Fig. 3-1. Here, the top scale displays electrical resistance in ohms. The basic scale is calibrated from 0 on the right, through 2K (2000 ohms), and to infinity (00) on the left. The resistance scale reads backward from the other scales; that is, its maximum value is at the left. The other scales increase to a maximum value at the right. When the multimeter's function/range switch (there may be a sep-

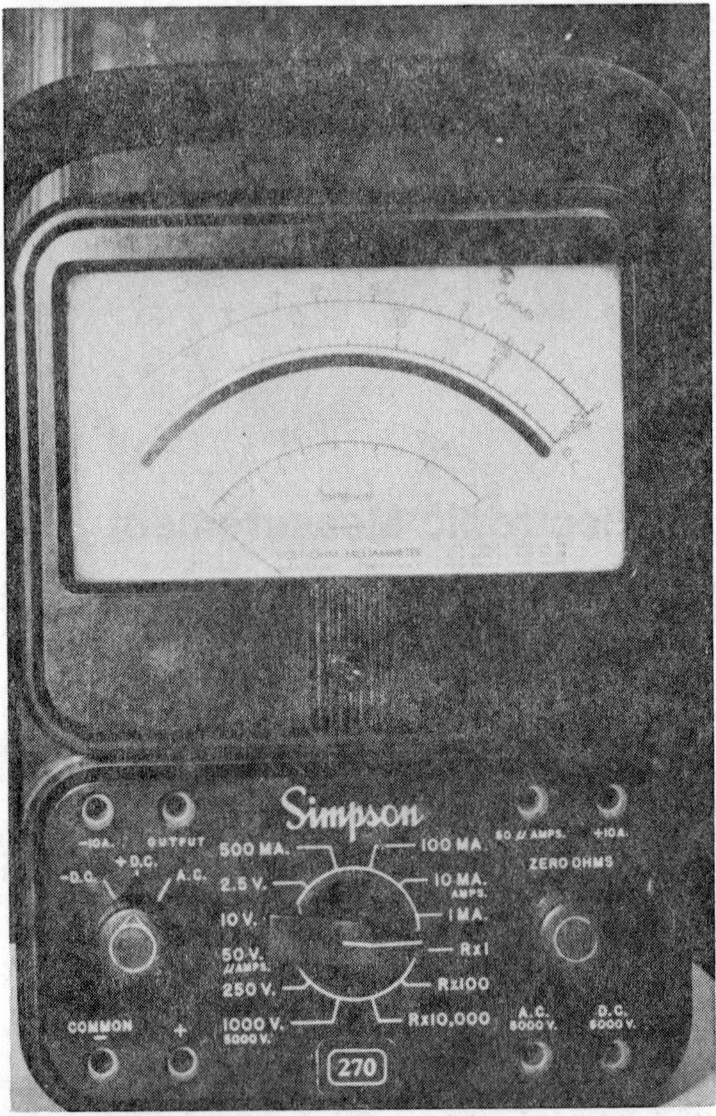

Fig. 3-1. A popular volt-ohmmeter or multimeter.

arate function and range switch, or they may be combined) is set to R × 1 (R stands for resistance), the value of a resistance is read on this scale. However, if the function/range switch is set to R × 10, the readings need to be multiplied by ten. The indication on the left side of the scale then becomes 20K or 20,000 ohms. VOMs are equipped with several resistance scales, which usually are R × 1, R × 10, R × 100, R × 1000 (or R × 1K), and R × 10K (10,000). (Other values may also appear.) Figure 3-2 shows the meter face of a VOM.

In the instruments shown in the last two figures, separate calibrated scales are provided for 0 to 10, 0 to 50, and 0 to 150 direct current, with the same scales provided for alternating current. These are used for voltage and current measurements. The range/function switch of a VOM may use these scales directly, or there may be some arithmetic necessary for these measurements. For example, if the range switch is set for 250 Vdc, use the 250 dc scale to read the voltage. If the instrument has a 25 V position on the range switch, use the 0 to 250 dc scale and mentally move the decimal point: 250 actually indicates 25.0 volts. This method can be applied to current measurements as well. For example, if the meter has a 10 A (ampere) position on the range switch, use the 0 to 10 dc scale and read the values directly. If it has a 50 mA (milliampere) range, use the 0 to 50 dc scale. If you select a 5 mA range, use the same scale and move the decimal point.

Figure 3-2 shows two additional scales, one of which is marked *2.5 Vac only*. This scale is read only when the range switch is set for 2.5 Vac. The other scale is marked *dB,* which stands for *decibels.* The decibel is a relative unit of electrical power that is frequently used in audio work. For this instrument, as indicated

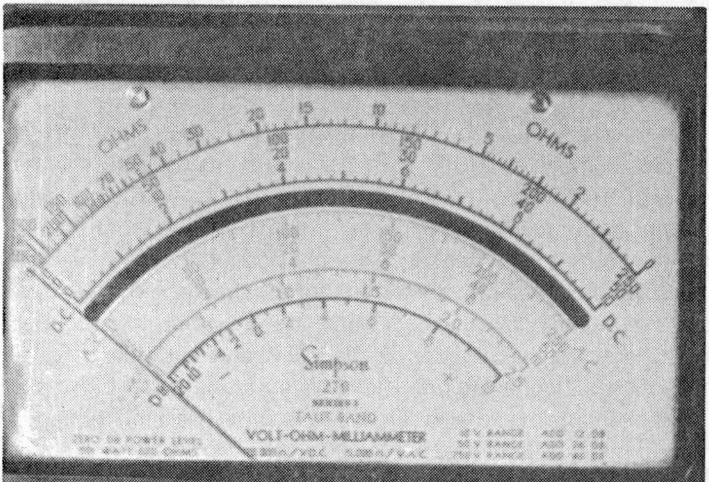

Fig. 3-2. The faceplate of a typical multimeter, showing the various measurement scales.

on the bottom of the face, zero dB is equivalent to a power level of 0.01 watt (W) measured across a 600-ohm load. This unit of measurement is used because the human ear has a response that is logarithmic to varying audio power levels. A change of one decibel in power level is barely detectable as a change in loudness by the human ear.

Some VOMs will have two jacks into which test leads can be plugged. Here, all selections of range and scale are made by means of the selector switch of the instrument. Others may have several jacks. On these units, it is very important that leads are plugged into the proper jacks when taking measurements.

The accuracy of multimeters can vary, depending upon the quality or tolerance of the meter movement and the multiplying resistors that are used to select the various measurement scales. Most price differences are based on the difference in cost between standard and precision components. An inexpensive multimeter can provide readings which are as little as 2 percent off, or as great as 10 percent off. A multimeter which is used by a service technician (and probably has a much higher price tag) may have an accuracy of 1.5 percent or 2 percent in dc ranges and 3 percent to 5 percent on ac ranges. This is necessary in this type of work, but for the projects in this book, none of the measurements are extremely critical and an inexpensive multimeter will work just fine.

Another type of multimeter which is now available is the *digital multimeter* (DMM). The major difference between this and the multimeter we have been discussing is that the DMM indicates the values on a multi-digit LED (light-emitting diode which glows red) or an LCD (liquid-crystal display which produces black numbers on an unlighted, shiny background). These types of meters can cost as much as $150, while a standard analog VOM can be had for about $30. Although the DMM provides both easier and more accurate readings, it is usually not worth the extra cost to the beginning experimenter on a tight budget.

Yet another version of the standard multimeter is the *vacuum tube voltmeter* (VTVM). The VTVM often has a greater number of ac and dc voltage ranges than the VOM, and a number of resistance ranges, but has no provision for measuring current. Its principle advantage over the analog VOM is that it has a much higher input impedance and does not load down certain sensitive circuits when voltages are being measured. The VTVM, however, does have the disadvantage of containing tubes, which must be allowed to warm up before using the instrument to take a measurement.

MULTIMETER PRECAUTIONS

When you use a multimeter, it's very important to select the appropriate range, particularly when you're using the analog VOM. Let's assume, for example, that a particular circuit operates from a 9-volt transistor battery. In this case, you can be fairly certain that the 10-volt range of the meter can be used safely. If you don't know the maximum voltage or current of a circuit, the multimeter should be set to the highest range for that function. Once an indication is seen on the meter, you can deduce what range the voltage or current is in and select the appropriate scale.

If you set the voltmeter range to 10 volts and hook the test probes into a circuit with 100 volts in it, the meter goes off scale and the needle may be damaged. This kind of damage is not covered by a warranty, and these instruments are quite expensive to be repaired. To avoid this, be sure you never apply the probes of a multimeter that is set to any ohms range to a circuit that has voltage in it. This can ruin the ohms scale and the meter movement as well.

Also, be careful how you connect the probes to a circuit when the meter is set to a current range (amperes, milliamperes, or microamperes). The meter acts like an almost perfect short circuit in the current range. If it is connected across a high-current battery or power supply, the meter will burn out. If it's connected across sensitive components, both the meter and the components would be destroyed. (Current measurements will be covered in greater detail later in this chapter.)

To ensure that measurements are accurate, it is sometimes necessary to make some simple adjustments to the meter. The meter movement in an analog VOM has a small screw head, which is located on the front panel, usually where the needle seems to end at the bottom of the meter movement. This is the *mechanical zero screw*. When you first purchase a VOM and periodically afterward, look carefully at the point where the needle rests when the instrument is disconnected from an external circuit. The needle should line up directly over the zero on any of the ac or dc scales (probably over the 00 on the ohms scale). If this is not the case with your meter, use a small screwdriver to gently turn the mechanical zero screw. Turning it counterclockwise usually moves the needle to the left, and clockwise usually moves it to the right. Turn the screw very carefully, as too much turning can damage the meter movement.

When using the ohms scale of the VOM, you'll need to adjust the control marked *Ohms Adjust* on the front panel of the meter. First, select the appropriate ohms range and short the test probes together. The meter needle probably comes to rest somewhere near the right of the meter face. Now, with the probes still shorted, turn the ohms adjust control until the meter reads exactly zero on the ohms scale. This calibrates the meter.

If you own a VTVM, you'll need to adjust the electrical zero control on the front panel. First, turn on the VTVM and allow it to warm up. Now, select the appropriate measurement range and short the probes together. Adjust the zero control until the meter indicates zero. To make resistance measurements with the VTVM, the high (infinity) end of the scale must also be set. Select the range, short the probes together, and adjust the zero control for a zero reading. Once this has been done, separate the probes and move the ohms adjust control until the needle rests directly over the 00 point on the scale. These two controls (zero and ohms adjust) interact somewhat, so it is necessary to repeat these two adjustments a few times until the meter reads 0 with the probes shorted and 00 with the probes separated.

RESISTANCE MEASUREMENTS

Resistance, which is measured in ohms, indicates the amount of current that will flow through a given material when a given voltage is applied. The lower the resistance, the greater the current for a given voltage. A circuit or component has a resistance of 1 ohm when an applied voltage of 1 volt causes a current of 1 ampere to flow through it. This is expressed by the three forms of Ohm's Law:

$$E = I \times R \qquad R = \frac{E}{I} \qquad I = \frac{E}{R}$$

In each case, E equals voltage in volts, I equals current in amperes, and R equals resistance in ohms. Even when small values are used (like milliamperes or microamperes), the expressions still hold true.

Figure 3-3 shows two resistors, each rated at 100 ohms. If the probes of the multimeter set in the ohms range are connected across either of these two resistors (after the ohms adjust is used), it will indicate a value of 100 ohms. However, when the two resistors are connected as shown in parallel, the total resistance will be less than the resistance of either of the two resistors; in this case, it will be 50 ohms. The reason is that each of the two

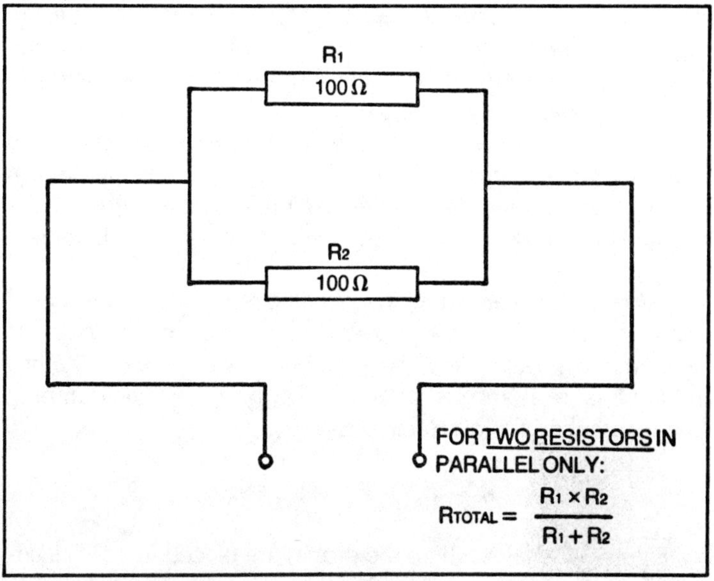

Fig. 3-3. Parallel connection of two resistors.

resistors offers the same amount of resistance to current flow, and the current is divided between them. The basic formula for determining the total resistance of any two resistors in parallel is:

$$R_{TOTAL} = \frac{R_1 \times R_2}{R_1 + R_2}$$

where R_1 and R_2 are the values of the two resistors in ohms.

Remember that when you're measuring resistance values in a circuit and the ohmmeter probes are connected across a resistor that is supposed to be valued at 100 ohms, there may be another component in parallel that results in a reading less than this value. In some cases, it may be necessary to disconnect one end of a resistor or another component in order to measure resistance. In other instances, you may wish to measure the total resistance value at a point in the circuit.

Figure 3-4 shows three resistors combined in a parallel circuit. To calculate the total value here, the following formula is used:

$$R_{TOTAL} = \frac{1}{\frac{1}{R_1} + \frac{1}{R_2} + \frac{1}{R_3}}$$

where R_1, R_2, and R_3 are the values of the three resistors. This formula can also be applied in a circuit with two resistors (by dropping R_3), or for a circuit with any number of resistors wired in parallel (by adding R_4, R_5, R_6, etc.).

Figure 3-5 shows how the multimeter probes are placed to measure a circuit composed of parallel resistances. The probes can be placed across any resistor in the circuit, with one probe on one lead and one probe on the other. The indicated value will remain the same.

When resistors are wired in series, the resistance values should be added. Figure 3-6 shows a circuit which contains four resistors combined in series. Here, each resistor has a value of 10 ohms and the total resistance is 40 ohms. The formula for determining total resistance in a series circuit is:

$$R = R_1 + R_2 + R_3, \text{ etc.}$$

Figure 3-7 shows where the probes are placed in a series circuit in order to measure total resistance. Also shown is the proper

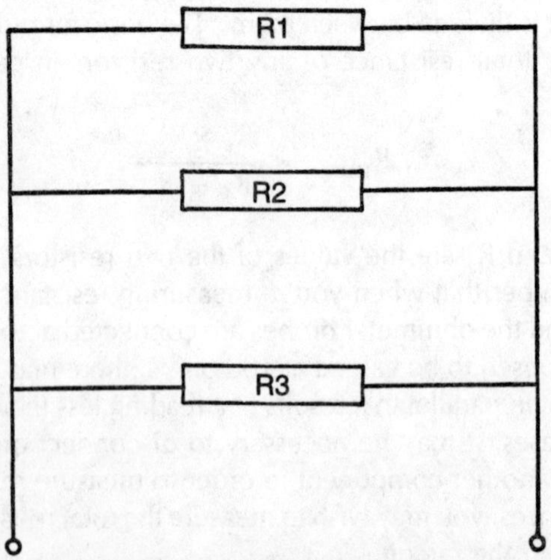

FOR ANY NUMBER OF RESISTORS IN PARALLEL:

$$R_{TOTAL} = \frac{1}{\frac{1}{R_1} + \frac{1}{R_2} + \frac{1}{R_3} + \cdots + \frac{1}{R_n}}$$

IN THIS CASE

$$R_{TOTAL} = \frac{1}{\frac{1}{R_1} + \frac{1}{R_2} + \frac{1}{R_3}}$$

Fig. 3-4. Three resistors connected in parallel.

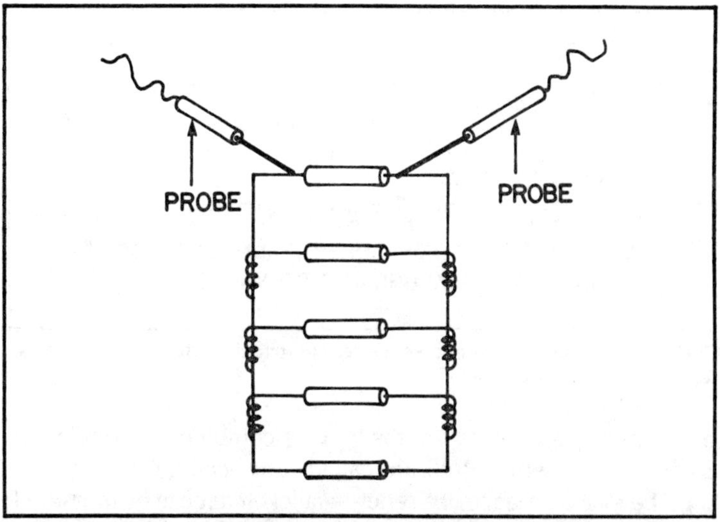

Fig. 3-5. Multimeter probe placement for measurement of parallel resistances.

placement for measuring a single resistor in a series string. Here, you can place the probes in such a manner as to measure any number of resistors in a series string without removing them from the circuit, as was required in a parallel configuration.

Figure 3-8 shows another type of arrangement. Here, the resistors are combined in a series-parallel circuit. This is a more complex configuration and is made up of two separate circuits, one consisting of two resistors in parallel and the other of two resistors in series. The two are then connected in a series-parallel combination. Here, total resistance is figured by treating each circuit separately. The total resistance of each is first determined, after which you must note the manner in which the two circuits have been combined (series or parallel). Figure 3-9 shows the two possi-

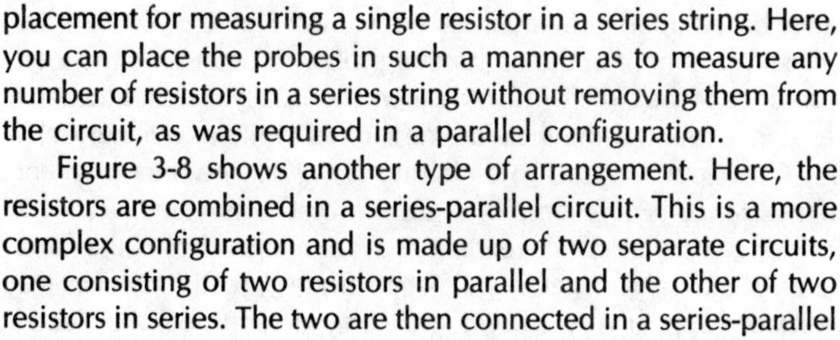

Fig. 3-6. A circuit consisting of four resistors connected in series.

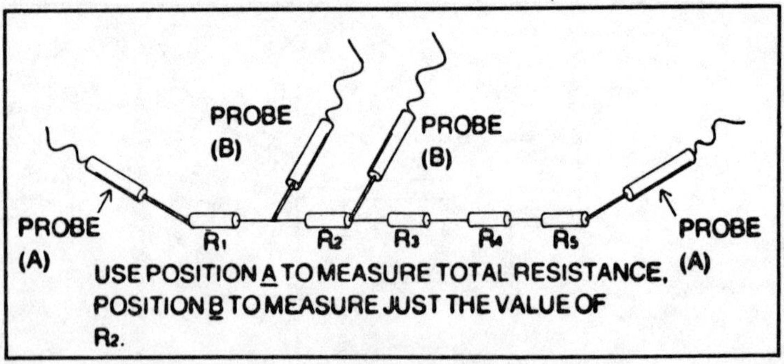

Fig. 3-7. Probe placement in a series resistor circuit to measure total resistance of all components.

ble combinations. In Part A, the two are combined in parallel; in B, they are in series. Both are called series-parallel circuits.

The total resistance of a series-parallel circuit can be measured quickly and accurately with a multimeter, which eliminates the need for mathematical computations to come up with total resistance based upon individual resistor values. Figure 3-9 shows the proper placement of the ohmmeter probes to determine the total resistance of the circuit. The measurement usually matches the total resistance found using the aforementioned formulas quite closely.

Other components in a circuit may affect the measurement of resistance. For example, a transformer connected across a resistor gives a resistance reading near zero, regardless of the rated value of the transformer. A solid-state diode or rectifier makes the resistance appear very low when the ohmmeter probes are placed across the component leads in one direction. The reading is closer to the actual value of the resistor when the ohmmeter probes are reversed. If you want to know the value of a single component

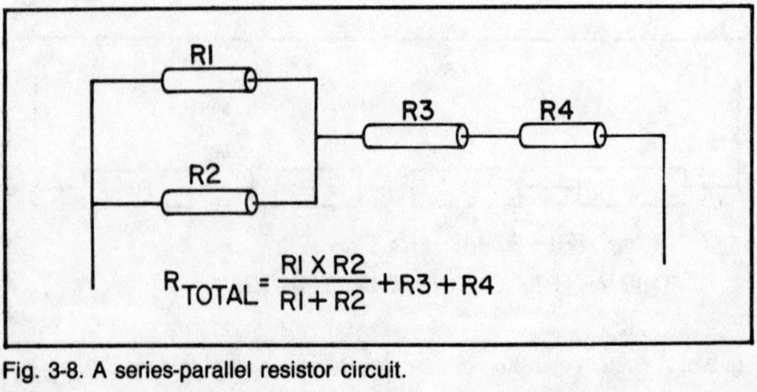

Fig. 3-8. A series-parallel resistor circuit.

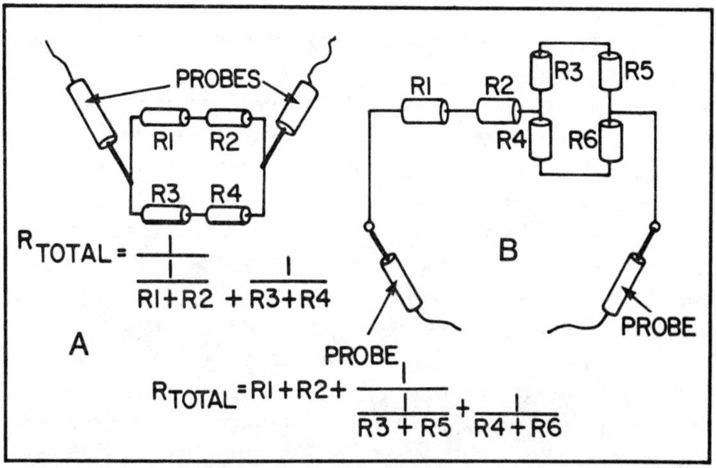

Fig. 3-9. Series-parallel combination resistance measurement. (A) Two circuits are combined in parallel. (B) Two circuits are combined in series.

or single circuit inside a complex circuit, it may be necessary to isolate the circuit.

An ohmmeter can be a very useful device for checking to see whether solid-state rectifiers or diodes are good or bad. Place the ohmmeter probes across the component leads of the diode. The reading will be either very high or very low (usually less than 10 ohms, but sometimes a bit higher for the low region). By reversing the black and red probes from one lead to the other, you will obtain an opposite reading if the diode is good (very low if the first reading was very high, or very high if the first reading was very low).

VOLTAGE MEASUREMENTS

When you use a multimeter to obtain voltage measurements, change the function/range switch to the appropriate position for dc measurements. The black lead or probe is the negative or ground connection, and the red lead or probe is the positive connection. Many types of electronic equipment are constructed on a metal chassis, which serves as the ground connection. Place the negative probe across the chassis and connect the positive probe to various voltage points to obtain readings. (You should be aware, however, that some electronic circuits have a positive ground; in this case, all other voltages points will be negative with respect to the chassis.)

Figure 3-10 shows a circuit that derives its power from a 9-volt battery. To take a voltage measurement, the negative probe of the ohmmeter is placed across the negative terminal of the battery, which is connected to circuit ground. The voltmeter scale is set in the appropriate range (the 10 or 15-volt dc scale is best). It may be necessary to decrease the scale range down to 5 volts or so as various points of the circuit are probed, because some of the voltages will be dropped in value by various resistors in the circuit. It is reasonable to assume that no potential which is higher than the battery voltage will be encountered if this is the only source of power. Now that the negative probe is connected and the meter range has been set, it is a simple matter to probe various parts of the circuit for indications of voltage. A dc voltmeter exhibits a very high resistance between its probes. Thus, it has very little effect on the circuit to which it is connected. A voltmeter has a very high internal resistance and is the equivalent of a very large value resistor being connected across the same two terminal.

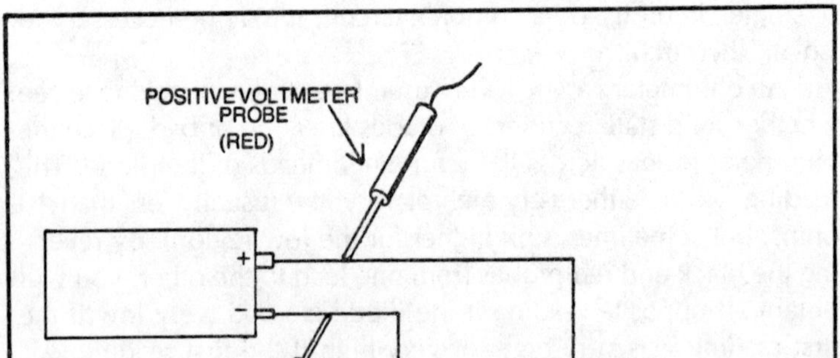

Fig. 3-10. An electronic circuit which derives its power from a 9-volt battery.

Only a tiny amount of current will flow. In a voltmeter circuit, this current is sampled and converted to a voltage reading on the meter scale. In most circuits, it isn't possible to cause damage when taking voltage measurements, unless you short-circuit components by allowing the metal tip of the probe to accidentally touch two points at the same time.

Caution: When measuring high voltage circuits (100 volts or more), always connect the negative probe so that it doesn't have to be held in your hand. Then connect the positive probe to an appropriate point in the circuit with one hand, and put the other hand in your pocket. If one hand were on the metal chassis (circuit ground) and the other came in contact with a high-voltage source, the full potential would pass through your heart by traveling through your arm and chest to complete the circuit. This type of severe electrical shock can cause your heart to stop beating, and if someone is not on hand to perform external heart massage, you could easily die. By keeping one hand in your pocket at all times when taking voltage measurements, any electrical shock received will be limited to one side of the body and won't travel through the heart. Using this method can greatly reduce the chance of a dangerous accidental shock.

When you probe different voltage sources within a piece of electronic equipment, the values will vary. The internal resistances of various components in a circuit are designed to reduce voltage to a level that is usable by other components. Many circuits contain both ac and dc voltages in the same conductors or devices. These voltages can sometimes interfere with the readings obtained on a multimeter, especially if the ac voltage is of a very high frequency. Most meters read the average voltage, as opposed to the peak ac value, which is considerably higher. Standard house current is said to be 110, 115, or 120 volts, depending on the area in which you live. This is the average value, sometimes called the RMS (root-mean square) value. The average value is normally used for performing power calculations, and this is the value which is usually read by ac voltmeters. The ac voltage ranges may follow the same pattern as the range of dc voltage, or it may be completely different, depending upon the meter. This mode of operation is designed to read ac voltage only and does not give accurate dc voltage readings.

It is not possible to take accurate ac voltage measurements with a VOM in a circuit that also has dc voltage present because the ac voltmeter circuit in the VOM also responds to dc, although

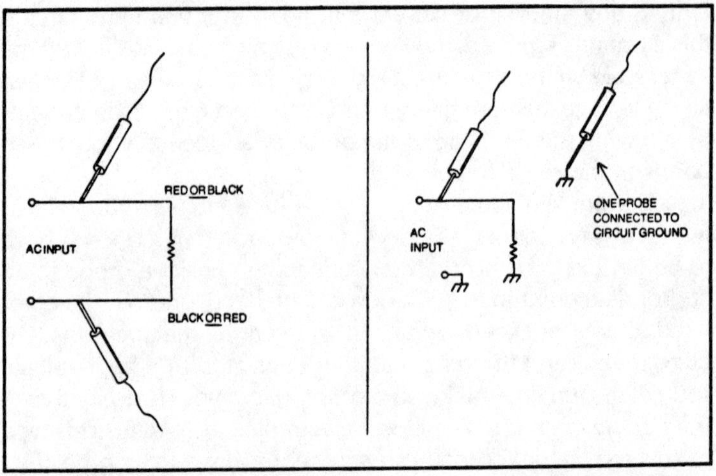

Fig. 3-11. Connection of a multimeter in order to take voltage measurements.

it is not calibrated. Thus, the voltage you read in this circuit is a combination of ac and dc.

One means of solving this problem is to connect a capacitor (0.1 microfarad should be about right—make sure it has a voltage rating higher than the combination of ac and dc in the circuit) in series with the voltmeter lead and the point in the circuit to be measured. This blocking capacitor blocks the dc voltage from the meter circuit and permits the ac to pass.

In the VOM pictured earlier in this chapter, a built-in blocking capacitor (600 V maximum rating) is available which is brought out to the front panel at the *output* jack. Here, all that's required is to connect the red test lead to *output* rather than +, set the meter to the appropriate ac voltage range, and read the scale in the normal manner. Some VOMs have this feature; if your doesn't, you can always use your own blocking capacitor in situations such as these. Figure 3-11 shows some of the ways a multimeter can be connected in a circuit to take voltage measurements.

DECIBELS

The multimeter pictured earlier in this chapter has a bottom scale on the meter face which is marked *dB* (for *decibels*). At the bottom of the meter, there is a legend which explains "ZERO dB POWER LEVEL .001 WATT 600 OHMS" and a table that says "10 V RANGE ADD 12 dB," etc.

The decibel is a unit for specifying relative power levels with respect to some reference power level. The commonly accepted reference level is 1 milliwatt (.001W), and levels stated with respect to this level are often called *dBm* for *dB with respect to one milliwatt*. Thus, −30 dBm means a level of +40 dB referred to 1 mW, which is 1 microwatt; or +40 dBm means a level of +40 dB referred to 1 mW, which is 10 watts. Decibels refer to power ratios expressed in logarithmic terms—and 10 dB represents a ratio of 10:1, 20 dB is 100:1, 30 dB is 1,000:1, and so on. This is only meaningful when both power levels used to produce this ratio are measured across the same impedance. In the case of the meter, the impedance is 600 ohms, which was selected because the telephone company has traditionally used standard 600 ohms impedance on their audio lines.

If you use the dB scale on a multimeter, set the meter to ac volts and the 2.5 V scale. If you are measuring across a 600-ohm impedance, this allows you to read decibels directly on the front panel scale. If the level is too high, switch to the 10 V scale and add 12 dB to the reading.

CURRENT MEASUREMENTS

The multimeter can be used to measure direct current and sometimes alternating current. In order to do this, it is sometimes necessary to move one or both of the test probes to a current jack on the front panel, although some multimeters perform this changeover by means of the mode switch. It is also necessary to choose the scale best suited for the anticipated range of current to be measured. If the current limit indicated by the range switch is exceeded, you can burn out a shunt element or even destroy the meter.

Most multimeters offer a 0 to 100 microampere (μA) scale, a 0 to 1 millampere (mA) scale, and a 0 to 1000 milliampere scale. This is the normal range; yours may vary. Many multimeters also have a high current scale that can read from 0 to 5 or 0 to 10 amperes.

Before actually taking any measurements, you'll need to know a bit more about the circuit to be tested. A schematic diagram is very useful as you measure the different current demands at various portions of the circuit.

The circuit must be broken at the point where the reading is to be taken in order to insert the ammeter in series. The ammeter then becomes a part of the circuit. Its reading increases as current demand increases, and vice versa. Figure 3-12 shows the proper placement for an ammeter when a reading of the total current drain of a circuit is desired. Here, the break in the circuit is made at the power supply input. The positive probe of the multimeter is connected to the positive terminal of the battery. The negative probe is connected to the circuit where the positive battery contact would normally connect. When the power switch on the equipment is switched on, current flows from the battery and through the meter, and an indication is given of the amount of current the circuit is drawing in normal operation.

The negative terminal of the battery may also be separated from its connection to the circuit; the positive probe of the multimeter is connected at this point. Then, the negative probe of the meter is connected to the remainder of the circuit. Either location is perfectly acceptable, and the placement of a meter in a series circuit is not critical. It will give the same measurement of current at any point in the series circuit.

If you meter has an ac current scale, it may be connected in exactly the same manner as the dc ammeter was, except that the power source cannot be a battery; instead, it must be some source of alternating current. Since an ac ammeter is a nonpolarized device, the polarity of the meter probes doesn't have to be considered.

The primary consideration when connecting an ammeter to a circuit is to be sure the proper connections are made regarding polarity and that the meter is always installed in series with the circuit whose current is to be measured. Keep in mind that a parallel connection in a circuit can cause meter damage and possibly blow fuses as well as cause circuit damage.

An ammeter can be an important device in determining how a circuit is operating. Some circuits may appear to be operating properly, but may actually be drawing too much current due to a malfunctioning component which may eventually cause com-

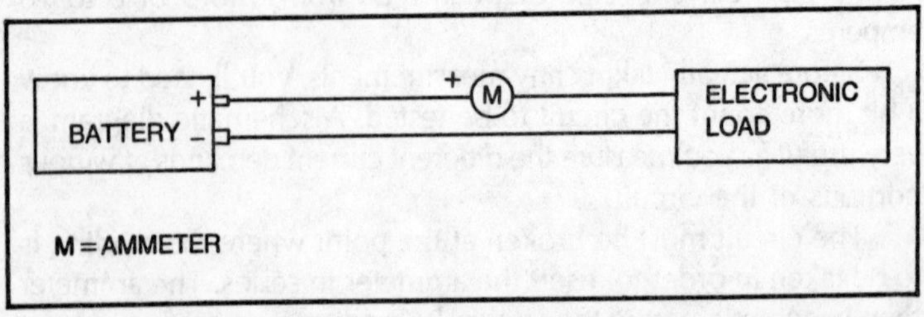

Fig. 3-12. Proper placement of an ammeter when reading the total current consumption of a circuit.

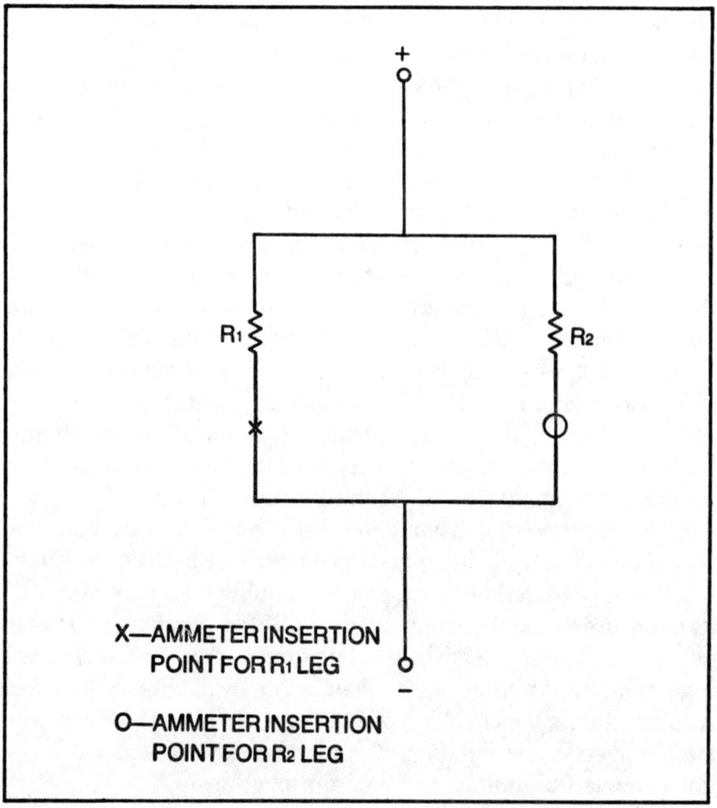

Fig. 3-13. Measuring current in one leg of a parallel resistor circuit.

plete circuit failure. Where the normal current drain is known, a multimeter used as an ammeter can be instrumental in detecting excessive current drain.

In parallel circuits which are all fed from the same power supply, an ammeter may be placed in one of the parallel legs to read the amount of current which is being drawn in that portion of the circuit. Figure 3-13 shows a circuit which consists of two resistors in parallel and indicates the connection points for the probes from the multimeter to measure current in that particular leg. To measure the overall current of this circuit, the meter would be placed at the power supply input point. By knowing the circuit and knowing which leg is to be measured, you can measure all types of current paths throughout the most complex electronic circuit.

Circuits which contain both alternating and direct current cannot be measured using a dc ammeter unless a particular leg of the circuit is measured at a point which is not subjected to the alternating current. For instance, in a radio frequency amplifier circuit, dc is fed through an rf choke, which is a device that prevents the flow of high frequencies in a circuit. At the plate connection of the amplifier tube, the direct current provided by the power supply and the alternating current generated by the amplifier is present. You cannot insert the ammeter at this point, because both ac and dc are flowing within the circuit. Instead, measure the direct current at a point in the circuit between the power supply and the rf choke, because at any point past the rf choke, alternating current at radio frequency is present along with the direct current.

The multimeter is an invaluable instrument for many electronic uses. It is essential for the electronics bench and takes the place of many discrete instruments. The multimeter allows the home builder to test and align many types of electronic circuits. As a troubleshooting aid, it really can't be beat.

The multimeter is usually the first electronic instrument obtained by home experimenters and electronics technicians alike. It is the most frequently used piece of equipment in the shop, so its many uses should be fully understood. The purchase of a good quality multimeter is an excellent investment. A poor quality meter may actually do more harm than good by giving misleading readings during the checking of electronic circuits. As long as you are involved in the building of electronic circuits, you will continue to use the multimeter over and over again.

Chapter 4

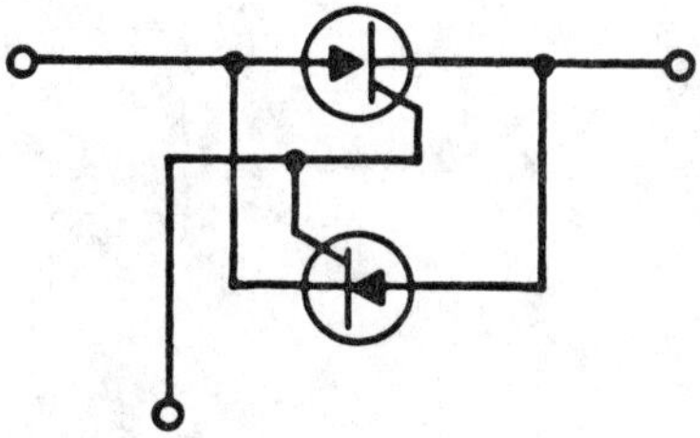

Electronic Project Building

MANY PEOPLE SHY AWAY FROM ELECTRONIC PROJECT BUILDING BEcause they feel that it is far too complicated, requires a great deal of expertise, and costs more money than they can afford. In fact, building your own circuits is relatively simple, as long as certain procedures are learned and put into practice from the onset. Safety is another concern of the beginning experimenter. Again, if you follow a few simple rules when working with electronic circuits, your pursuit will be far less dangerous than driving an automobile.

The cost factor involved in building electronic projects is not as great as many people think. Even if you are on a limited budget, it is relatively easy to obtain all tools and materials needed, although you may have to shop around for the best bargains. But many bargains are available, if you are willing to take the time to find them.

This chapter introduces you to the art of building electronic projects, and provides an explanation and description of all the tools you'll need to start out. The more experienced builder can use this chapter as a refresher course.

Remember to take each phase of circuit building slowly and carefully, making sure that everything is as it should be before proceeding to the next step. Start with the more simple projects in this book before going on to those which are more complex. None of the projects included here is extremely complicated, and the

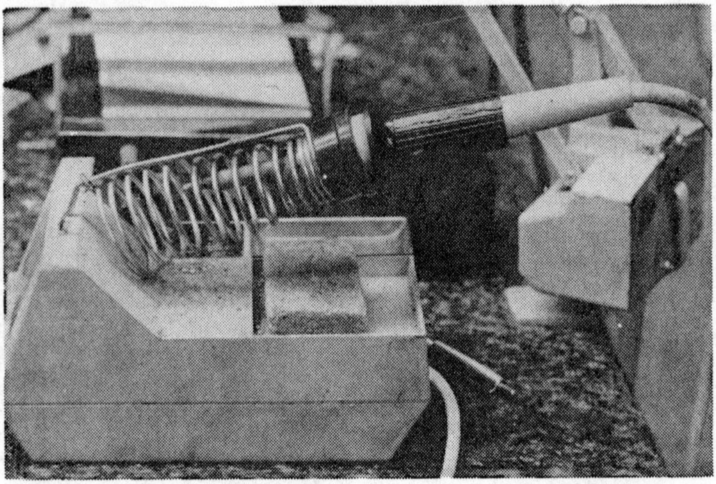

Fig. 4-1. A popular soldering station.

average person with a bit of working knowledge of electronics should have few problems.

TOOLS AND ACCESSORIES

Your electronics workbench should be equipped with all the tools you'll need. One of the most important is a soldering iron. These are available from many different manufacturers in a wide variety of configurations. The most common type is called a *soldering pencil*, which usually has a power rating of 25 to 30 watts. Because of the small size of solid-state components, the soldering pencil is recommended for most of the projects in this book; it provides adequate heat, but not so much that components are in danger of being destroyed. You can buy one at most hobby stores and in mail order catalogs and they sell for about $10 or $20.

Another type of device is called the *soldering station*, shown in Fig. 4-1, which consists of pencil soldering iron, an insulated holder, and a control box. The control box maintains a steady temperature at the tip of the pencil at all times. When the temperature dips below a certain level, the device turns on. Likewise, it shuts down when the temperature rises above a specified level. Although this provides a great deal of convenience, a soldering station costs a bit more money. It is not an absolute necessity for the beginner on a tight budget.

You will also need a supply a solder for use with the soldering iron. The type used in electronic project building is called *rosin core solder*. It is always identified as such and is available from most electronic hobby stores. Rosin core solder should never be confused with *acid core solder*, which is sold in hardware stores. Acid core solder is used primarily in plumbing and contains a corrosive acid which quickly destroys electronic circuits and their components.

Other tools which will be used in building electronic projects include:

- a pair of needlenose pliers
- diagonal wire cutters
- a small pocket knife
- a wire stripper
- an assortment of screwdrivers (both Phillips and flathead)
- nutdrivers
- an assortment of nuts and bolts
- electrical tape
- alligator clips
- clip-on heatsinks

For most of these items, you do not need to buy the most expensive or highest quality available; neither should you purchase the bargain-basement variety. Since you'll use the diagonal wire cutters and the needlenose pliers most often, you may wish to purchase top-quality ones, but medium-priced tools will suffice for the others. Diagonal wire cutters are used to cut away excess lead lengths, remove components from circuit boards, and cut away sections of the perforated circuit board after the circuit is complete. Needlenose pliers are used to bend wire leads and wrap them in order to form a strong mechanical connection before soldering. (This is explained in detail later in this chapter.)

These tools should be adequate for all the circuits in this book. More specialized projects require the use of additional tools, and if you're more experienced in the art of electronic project building you'll probably have many more tools which have been purchased over a long period of time as the need arose for them. If you're a beginner, start off with the basics, and build up your tool complement as needed.

SOLDERING TECHNIQUE

Soldering is one of the most important skills used in electronic project building, regardless of the nature of the project and the components used. It also creates the greatest number of problems, for beginner and experienced experimenter alike. This is not because soldering is extremely complex and requires years to learn properly; rather, it requires only great care and strict adherence to a few simple rules. A poor solder connection at a single point in a circuit can make the entire circuit inoperational or erratic, and troubleshooting can become a lengthy process of resoldering each and every connection in order to correct the problem. This type of situation can be easily avoided by taking the time to make each solder connection carefully according to the following rules:

- Carefully inspect all conductors which are to be soldered to make certain they are free from oil, dirt, or insulation.
- Carefully scrape away any of these materials until the surfaces are clean and shiny.
- Wrap the conductors to be soldered tightly. This creates a good mechanical joint, or bond. Check to ensure the firmness of this wrapping procedure by gently tugging at the component leads. If no movement is observed, a firm mechanical joint has been made.
- Turn on the soldering iron and allow it to heat to the proper temperature. Never attempt to use a soldering iron before it has warmed up sufficiently.
- Once the soldering iron is at the right temperature, apply the tip to the joint, not to the solder. This is probably the most crucial step in soldering, and if it's not done properly, a cold solder joint results. A cold solder joint is most often caused by dropping a glob of solder onto the joint before the component leads or wires have been heated by the tip of the soldering iron. After the joint has been heated by the iron, place a small amount of solder against it and allow the solder to move freely around the conductors. When the joint has been heated to the same temperature as the soldering iron, the solder melts and flows into every part of the wrapped conductors. Be sure to use only a small amount of solder. Too much solder can also create a cold solder joint, because an unusually large amount of solder is probably not completely melted on the inside.
- Once the solder is flowing in the joint, remove the tip of the soldering iron and allow about 20 seconds for the solder to cool. During this period of time, make certain that the conductors are not allowed to move, as this may cause the solder to crack

or become loose. Once the 20 second period has elapsed, gently tug on the conductors to make certain no physical movement occurs where the bond has been made. Now closely examine the joint, looking for any signs of a cold solder joint, such as a dull surface or large globs of solder. A good solder joint has a shiny smooth appearance.

These rules are quite simple, and if followed carefully on each and every solder connection, you should have very few problems in this area with your electronic projects. It is a good idea to practice this procedure on a scrap section of perforated circuit board with some junk box components until the whole process becomes second nature. Some types of components, especially solid-state devices, can be damaged by the heat of even a low-wattage soldering pencil if they get too hot. Thus, speed can be quite important when soldering to ensure that no components are heated past their point of endurance. A few hours of practice should provide you with the skill needed to make proper solder joints in a minimum amount of time.

The *heatsink* is quite useful during soldering. It sinks (absorbs) heat and dissipates it into the surrounding air, away from delicate components. An alligator clip can be used as a heatsink; it's less expensive and will do the job just as well. Alternately, a pair of needlenose pliers can be used. These are placed at a point near the shell of the component, as shown in Fig. 4-2. Some of the heat travels up the lead from the point where it is being soldered, but the greater portion is absorbed by the needlenose pliers. The only disadvantage of this tool is that it requires that you hold it in place with one hand. If you use alligator clips or heatsinks, you clip them in place, leaving both your hands free for other procedures. Whatever you use, place it close to the case of the component. If the heatsink is placed too close to the joint, it draws some of the heat away from the joint, creating a cold solder joint.

CIRCUIT BOARD MOUNTING METHODS

Most of the projects in this book are designed to be constructed on perforated circuit board. This can be purchased from any electronics hobby store and provides a great deal of versatility in building many different types of projects. It is quite inexpensive and can be purchased in single sheets or in packages containing

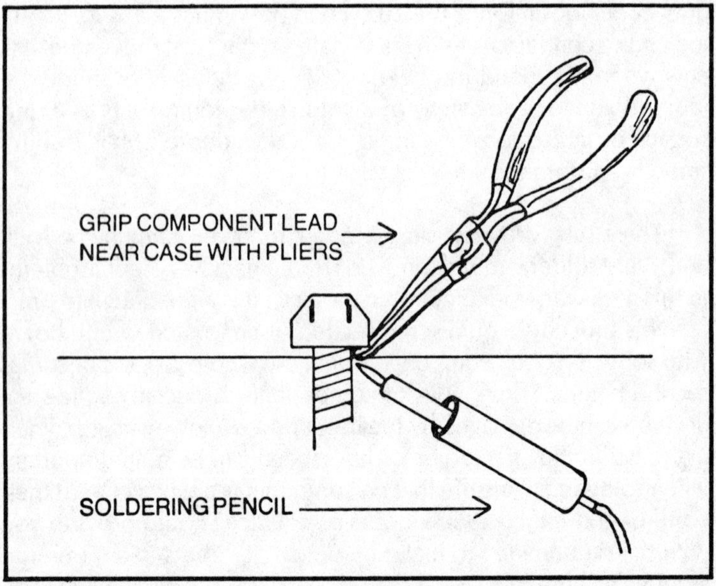

Fig. 4-2. Using needlenose pliers as a heatsink when soldering a solid-state component.

a variety of different sizes of sheets. It is made up of a rigid dielectric or insulating strip, through which rows of holes have been punched to allow for the insertion of component leads and hookup wires. A piece of perforated circuit board is shown in Fig. 4-3. Some types of perfboard will have closely-spaced holes in greater quan-

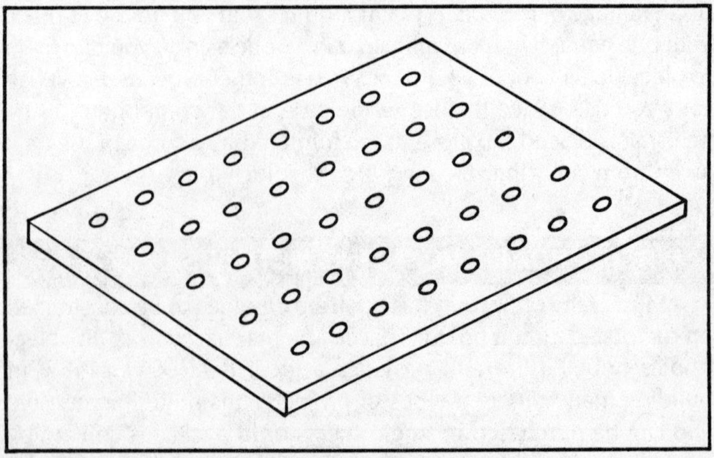

Fig. 4-3. A small section of perforated circuit board.

tity, while others will contain holes which are more widely spaced and less numerous. Here, you must decide before beginning to assemble a project which type is best, depending upon the number of components, their size, and the mounting method chosen. Perforated circuit board is also available in a number of thicknesses, although most types weigh less than an ounce or so. Thicker boards are used when heavy components are incorporated into a project, ensuring the proper amount of support.

There are two established methods of mounting components on perforated circuit board. Both involve pushing the component leads through the holes in the board and making all solder connections at the rear. The more common method of the two is known as *horizontal mounting* and is shown in Fig. 4-4. All components are installed lengthwise on the circuit board, with the leads pushed through the holes which are spaced the equivalent of the component's physical length. This method is recommended for beginners, because it is easier to distinguish individual circuit sections. It requires more space than the other method of mounting.

Figure 4-5 shows an example of vertical mounting, in which each component is mounted vertically in relation to the surface of the perforated circuit board. Even though all leads are still pushed through the holes in the board, the holes can be spaced only a fraction of the distance apart. Here, the vertical space above the board is utilized, and the circuit board can thus be much smaller. If vertical mounting is used, it is necessary to insulate the leads which exit the top of the compartment and descend along its side to the circuit board. This can be done with hollow rubber tubing, available in most hobby stores. Slip the tubing over the component lead before it is soldered into place so it completely covers the exposed lead from its point of exit from the case to the

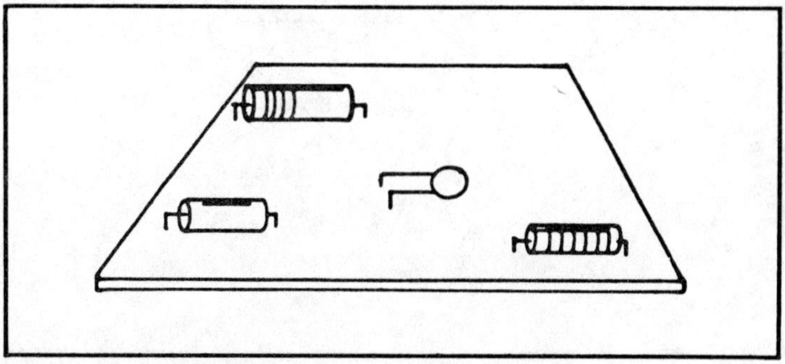

Fig. 4-4. Horizontal mounting of components on perfboard.

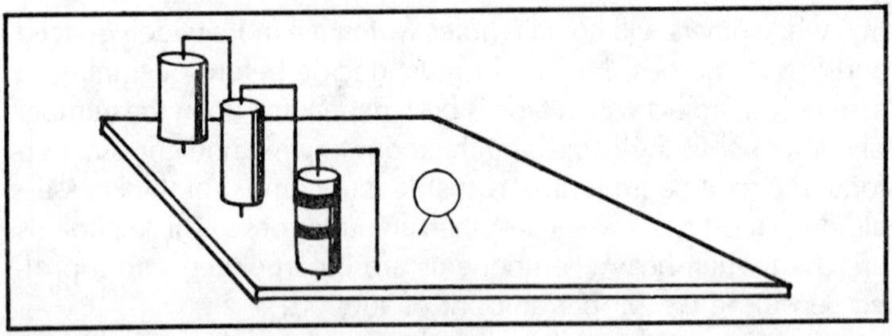

Fig. 4-5. Vertically mounted components.

entrance of the circuit board. This procedure is illustrated in Fig. 4-6.

If you choose vertical mounting, you must realize that only the circuit board size has been decreased—the space needed to house the components is simply transferred to the vertical plane. Sometimes, it is desirable or even necessary to combine both mounting methods in a single circuit. This may be quite handy when a half-completed project on a piece of perfboard is discovered not to have adequate space. Here, it is quite easy to switch to vertical mounting for some of the larger components in order to make everything fit. Regardless of which method is chosen, remember to follow good building practices through construction, and choose the mounting technique which suits the project best in both appearance and operation.

When mounting components on a section of perforated cir-

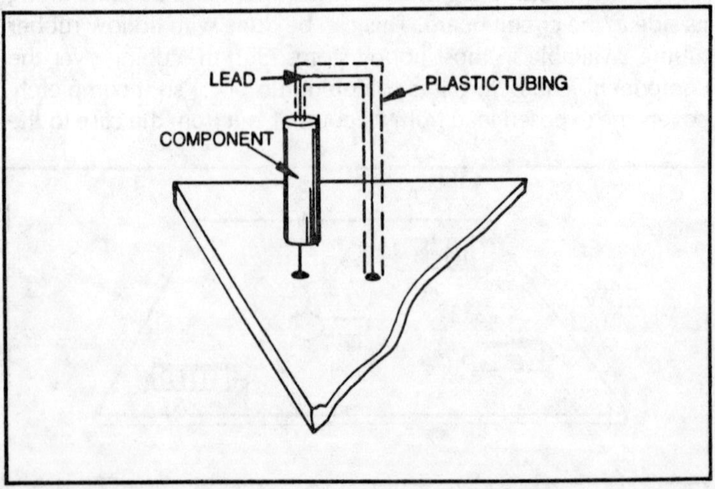

Fig. 4-6. Insulation of a component lead with rubber tubing.

cuit board, place components which will connect to each other to close proximity to the top of the board. This ensures good wiring technique, which mandates that wiring be kept as short as possible with a minimum amount of insulated hookup wire. Sometimes it is necessary to use short lengths of hookup wire in a circuit, but good planning before assembly can save both soldering time and space on the circuit board. Many of the projects in this book include a suggested parts layout. Of course, you can make modifications in the layout if you want to, as long as all of the wiring is kept as short as possible. It is important to realize that the simpler a circuit is, the more reliable it will be once completed and in operation.

THE ELECTRONICS WORKBENCH

Everyone who is involved in electronic project building has a different situation to deal with regarding a work area. No two homes are alike, and you may have to make some sacrifices to set a space where you can pursue this hobby. There are some criteria that should be considered when making a decision as to where you will be building your electronic projects.

An electronics workbench allows you to comfortably and conveniently build circuits with a minimum of interruptions. In some cases, this may be a kitchen table that is converted to a work area in the evening hours. Here, all that is required is that you provide adequate protection for the surface to ensure that it is not damaged by sharp edges of tools or components, and that no solder is allowed to drip onto the table. Also, be sure that the chair you will be using provides adequate support for your body, as you will sometimes be sitting in the same position for long periods of time. Lighting is quite important when building electronic projects, as you will be looking at some tiny components and trying to twist them together. A good light source makes the whole process much easier, and makes it possible to work for longer periods without eye strain. If an adequate light source is not already available in the area you have selected for building, you may want to purchase a second-hand lamp that will rest high enough on the table to provide adequate lighting over the entire work area. I also recommend that you lay your tools out in an orderly manner, rather than leave them on the floor in a toolbox. This makes it much easier to locate the tool you need quickly, instead of having to shift your position and rifle through a whole box of tools. This type of preparation saves time during assembly, and your concentration is not interrupted from the project at hand.

When building circuits on perforated circuit board, you will find that you are constantly turning the board one way and then the other as component leads are pushed through the holes and wrapped at the rear. I have devised a simple solution to this constant motion. When selecting a section of perfboard, I use a piece which is somewhat oversized for the circuit being built. This is then placed on some wooden blocks for support, as shown in Fig. 4-7. The leads are then inserted through the top and wrapped without ever having to move the board. If more stability is desired, the perfboard may be secured to the blocks by means of a few tacks. Once all components are mounted and you are ready to begin soldering, the tacks are removed, the board is turned over, and soldering can take place. After soldering is complete, diagonal cutters are used to trim away any excess circuit board. This method is both convenient and time-saving.

A vise works equally well. Insert the perfboard in the vise, and turn it around for soldering as described above.

One advantage of having a work area designated specifically for project construction is that you don't have to clear everything away when you stop for the day. Of course, this does not preclude providing some sort of protection for the circuit being constructed to ensure that nothing is exposed in any way. Place the circuit away from the edge of the bench, and make sure that no tools or heavy objects are nearby that could bang against it or fall on it from overhead. When a circuit is in an incomplete state and has not yet been installed in an enclosure, it can easily be damaged.

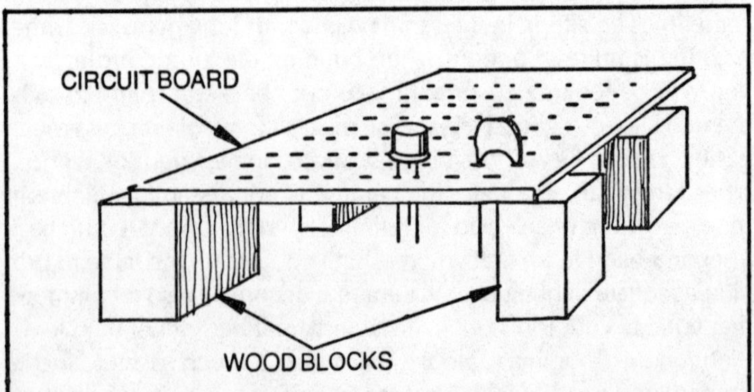

Fig. 4-7. Using wooden blocks to support circuit board during construction.

BUILDING HABITS

When working with electronics and electronic circuits, it is absolutely essential that you establish good working habits right from the start. Some of these suggestions may seem like simple common sense to you, but they do bear repeating. Ignoring them can result in improperly constructed circuits and possibly accidents.

One important practice is to never start assembly of a project until all components, connectors, and enclosure—everything you need for the project—is on hand. If you begin a project with some components missing, there will be an interruption in the order in which they are mounted, and this lack of order can make you prone to missing an important step at some point during assembly.

Another good building habit is to take frequent breaks during construction to rest both your mind and your body. Sitting in the same position for hours on end, staring at tiny leads and components, can take its toll on you both mentally and physically. Don't even attempt to work on a project when you are already fatigued. It's much easier to make mistakes when you're tired; and remember, you are dealing with electricity here, which can be potentially lethal, especially when your mind is not clear.

When you feel the least bit tired, stop what you're doing and take a ten or fifteen minute break until you're refreshed. Also, don't set a specific time for the completion of a project. If you find you are running behind schedule, you may begin to rush or work past your point of concentration. This often results in wiring errors, reversed polarities, and cold solder joints.

Building electronic circuits is not an extremely complex process, nor is it difficult for the beginner to learn. If you follow the directions provided in the projects chapter, and the common sense procedures outlined in this chapter, your circuits should work the first time around. A great many circuit problems can be easily avoided.

Chapter 5

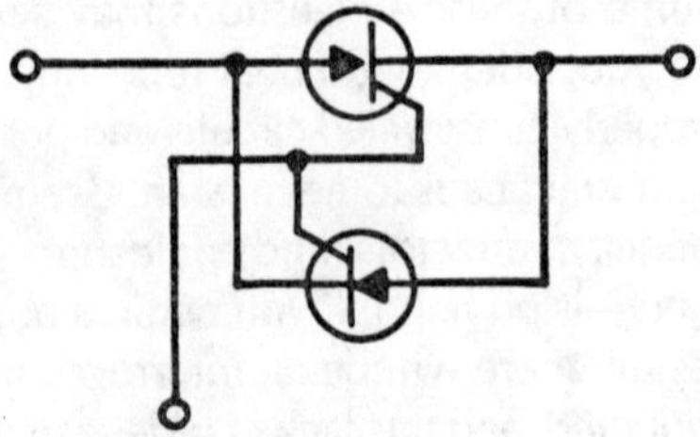

Obtaining and Referencing Components

MANY WOULD-BE ELECTRONIC ENTHUSIASTS ARE RELUCTANT TO delve into project building for the simple reason that they have no sources for obtaining many of the electronic components needed. Anyone who is interested in building electronic projects should be aware that there are a multitude of sources for electronic components. Most of the circuits provided in this book are made from relatively common components which can be locally obtained. Those which contain components which may be difficult to obtain are indicated as such, and a supplier is provided. Also, many parts can be directly replaced by equivalents from another manufacturer if you are unable to locate the part specified.

The place to start in locating electronic components and accessories is your local hobby store. Here, you should be able to find close to 90 percent of the items used to build most of the circuits in this book. These same stores also have available mail order catalogs from such companies as Allied Electronics, Fair Radio Sales, and other companies that have been suppliers of electronic components for many years. The latter company deals primarily in government surplus at very reasonable prices. Allied Electronics, on the other hand, is a major supplier of new components and devices, ranging from transistors, diodes, and photocells to electronic computers and industrial electronic equipment. The two companies mentioned here are among many who can provide you with electronic components, and tools.

A trip to a local electronics repair facility can quickly provide you with the names of other part suppliers which may be nearby. You may be able to order almost any electronic part imaginable through these sources. Once you have the name of the parts supplier, a phone call should get you a complete selection of catalogs from the manufacturers whose products they handle. Typically, you will find catalogs from such manufacturers as RCA, Motorola, Sprague, Sylvania, Miller, Amphenol, and General Electric, which will probably be sufficient to supply fully 99% of your needs. If this supplier is located in your town or city, this will probably mean same-day service if the parts needed are in stock. If you live in a rural area, there is a good chance that these suppliers have route men who pass through on a regular basis. A phone call to the supplier will usually mean that your order can be delivered within a few days. If you elect to go the mail order route, a phone call to their order desk (often a toll-free 800 number) will allow your order to be shipped out the same or following day.

The cost factor may drive a lot of experimenters away from electronic project building, and there is a great fluctuation in prices of equivalent components from different suppliers and dealers. For this reason, it is absolutely essential to shop around for the best deal. I have economically replaced solid-state components called for in a circuit with a component from another manufacturer. The two parts were electronically equivalent, but one was less than half the price of the other. This is where a lot of unnecessary expense may enter the electronic building hobby. This especially applies to solid-state devices such as diodes, integrated circuits, and transistors.

CROSS-REFERENCING

One of the most valuable aids which can be obtained from the various suppliers of electronic components is cross-reference material for the manufacturers who make solid-state devices. Each manufacturer prints their own catalog or replacement guide. These list almost every transistor, diode, and integrated circuit manufactured today, and tell you which devices from this particular manufacturer are replacements. It is a good idea to have replacement guides from as many manufacturers as possible. For this discussion, three solid-state component manufacturers have been chosen to demonstrate device cross-referencing: Sylvania, General Electric, and RCA.

As an example, let's assume that an electronic circuit calls for a 2N5060 silicon-controlled rectifier. The 2N designation is rarely seen anymore: it's the old generic name for solid-state devices. Today, most manufacturers have an individual and specialized numbering system for naming their devices. In search of a 2N5060 SCR, we first look through the Sylvania replacement guide. It reads just like a progression list. First, we locate the prefix 2N and work from there. Reading up the list, we have 2N5054, 2N5058, 2N5059, and finally, 2N5060. To the right of the 2N5060 reference is the Sylvania replacement. We find that an ECG 5400 is a direct replacement for the 2N5060. The same can be done using an RCA catalog, which tells us that the SK3950 is a direct replacement for the 2N5060. It is also a direct replacement for the ECG5400. The General Electric replacement guide specifies a GEMR-5 as the exact replacement. If the SCR part number supplied in the circuit diagram were that of another manufacturer, this device could also be referenced to the current device equivalent from another manufacturer. All of these devices are electronically equivalent. The great majority of them contain the same case and lead configurations, but it is wise to make certain the leads are arranged in the same manner. Sometimes, different cases are used to house equivalent chips, but this type of occurence is usually specified in the cross-reference manuals.

The vast number of solid-state part identifications often confuse and frustrate beginning and experienced experimenters alike. However, if you keep the cross-reference manuals close at hand at all times and use them properly, you will find that you sometimes have an equivalent device already on hand, either new or in a junk box circuit.

THE EXPERIMENTER'S JUNK BOX

Virtually every home experimenter tends to accumulate a "junk box" of components in a very short period of time. This term is not actually descriptive, since many of the components are simply being stored for use in future projects, and thus, are not really junk. Actually, a junk box can be any assortment of crates, boxes, and cabinets which are filled with surplus transmitters, receivers, old data circuit boards, electric clocks, etc. Through many surplus outlets, it's possible to obtain brand new circuit boards from discontinued lines. Manufacturers dump many electronic components and circuits into such industrial channels, and some very good buys can be had. It is not uncommon to purchased a transistor or IC from one of these mail order dealers guaranteed to be new and operational for one-tenth of the price of an equivalent component on the retail market. For instance, rectifier diodes may be purchased for 11 cents apiece which might cost $2.50 if purchased from a retail repair facility.

Many enterprising experimenters keep track of products through ads placed by distributors and dealers in electronics magazines. They also send letters to distributors, asking to be placed on a mailing list to receive constant updates on available products. In this manner, when a good deal which offers components of immediate or future use comes along, a purchase is made. These parts are then stored in the junk box until they are needed. Sometimes, a purchase is made which seemed to be attractive (due to price) without any idea of what the components may be used for. Intuition comes in quite handy here, and occasionally, these types of purchases yield a wealth of usage in the future. Often, however, the components sit around for a few years and are traded off to other experimenters, resold, or tossed in the real junk box (often called the trash can). In the early stages of building up a supply of components, it is quite easy to become enamored with quantity rather than quality. This can often become a wasteful and costly mistake. Don't become so parts-conscious that you buy everything available at a low price. Some components are priced cheaply because their practical uses are ridiculously few and far between.

Another source of junk box components is amateur radio operators, CB enthusiasts, and other more experienced experimenters who have allowed their junk boxes to become whole rooms of junk. Their amount of junk has reached the point where it is overflowing the owner's ability to keep it in an orderly fashion. Here, it may be possible to simply haul away wheelbarrow loads of components for little or nothing, as the owner may prefer this rather than having to pay to have it hauled to a junkyard. Local radio and television stations may be another possible source, as well as television and stereo repair facilities. These types of business sometimes have old television receivers around that contain a wealth of salvageable electronic components that can be salvaged. This is a tedious and dirty job, but it is sometimes worth the effort.

I have been quite successful in salvaging electronic components from printed circuit boards in a speedy manner by cutting the circuit board out from under its components. The only tools required are diagonal cutters and needlenose pliers. Most of the older types of circuit board material chip away in large pieces when cut. The green, glass-epoxy boards can present major problems with small cutters, and a larger tool should be brought into play. Usually, the entire circuit board can be cut away without having to use a soldering iron. Any solder remaining on component leads turns to dust when tightly squeezed in the jaws of long-nosed pliers. A medium-sized section of circuit board containing 100 components can be cleared in less than an hour. Using this method, you may find that some of the components are damaged, but if you have a special interest in any components on the circuit board, it is an easy matter to remove these ahead of time by desoldering their leads in the conventional manner or by using clip-on heatsinks to protect them.

Local electronics hobby stores can also be a source of junk box parts. These outlets often display defective or damaged merchandise in a special area, which can be had quite cheaply. In my home town, there is a Radio Shack store which regularly offers discontinued items for pennies on the dollar. Even if items are defective or damaged in some way, many of them can be repaired or salvaged for individual components.

It is not that difficult to obtain some of the more common mechanical parts needed for electronic project building either. These include such devices as switches, relays, and mechanical timers, which may often be put to use in electronic circuits. These can be found through some electronic outlets and in the junk boxes of other experimenters. Parts such as these are often quite expensive, and surplus or used components sometimes work just as well as the new counterpart. Government and industrial surplus components which may have originally cost hundreds or even thousands of dollars and can now be had for less than $10.00. This price differential is not very unusual in the field of surplus components, as you'll undoubtedly discover as you pursue these markets.

STORING ELECTRONIC COMPONENTS

Once you have built up an inventory of electronic components, you will need an efficient means of storing and keeping track of exactly what you have on hand for future needs. Obviously, the prime purpose of obtaining these components is to allow you more flexibility when building projects without having to order every part new. If you don't have a good storage system and everything is strewn all over your work area, you have defeated the whole purpose.

Fortunately, it is quite simple and inexpensive to categorize the majority of your surplus components, and the entire process need take no more than a weekend. It is absolutely essential to take the time to devise an orderly method of storing your junk box components.

It is best to start with the heavy components, such as large transformers and chokes. These can all be placed in a wooden or cardboard box which can be stored in a closed cabinet or on a shelf that is strong enough to hold the weight. Be sure to identify the contents of the box with a colored marker. If you have a large assortment of components in this category, it may be a good idea to divide them further into sub-categories. All of the low-voltage devices could be one category, the medium ones another, and so on, with each carton labeled as such.

Smaller boxes are required for smaller components, such as audio transformers, switches, and relays. Cigar boxes are ideal here. They are large enough to house these parts and usually fit easily on a shelf over the workbench. Usually, only one type of component is stored in each box, but you should be the judge of this method. If your present supply is small, this may not be necessary, but make sure each box is labeled clearly and can thus provide you with the information needed to locate a specific part. For example, one box might be labeled *AM radio parts* and might include capacitors, ferrite rod antennas, speakers, and so on, that have been removed from AM radios.

Individual components such as resistors, capacitors, and solid-state devices can be stored on workbench shelves with plastic trays. These can be purchased almost anywhere, and one such unit is shown in Fig. 5-1. This type of cabinet may contain twelve to forty plastic trays, which slide forward to provide access to their contents. Often, these drawers can be subdivided into different compartments with plastic spacers included with the cabinet. Different types of transistors, diodes, and capacitors can be stored in a single drawer while being kept apart from the others in the same drawer. Figure 5-2 shows a small parts cabinet which is especially designed for stacking purposes. Here, the bottom is recessed and can easily slip over the raised, mating surface found at the top of a similar unit, shown in Fig. 5-3. Figure 5-4 shows how several of these units can be stacked to conserve space in the work area.

Resistors are usually sorted and stored according to value. Fig-

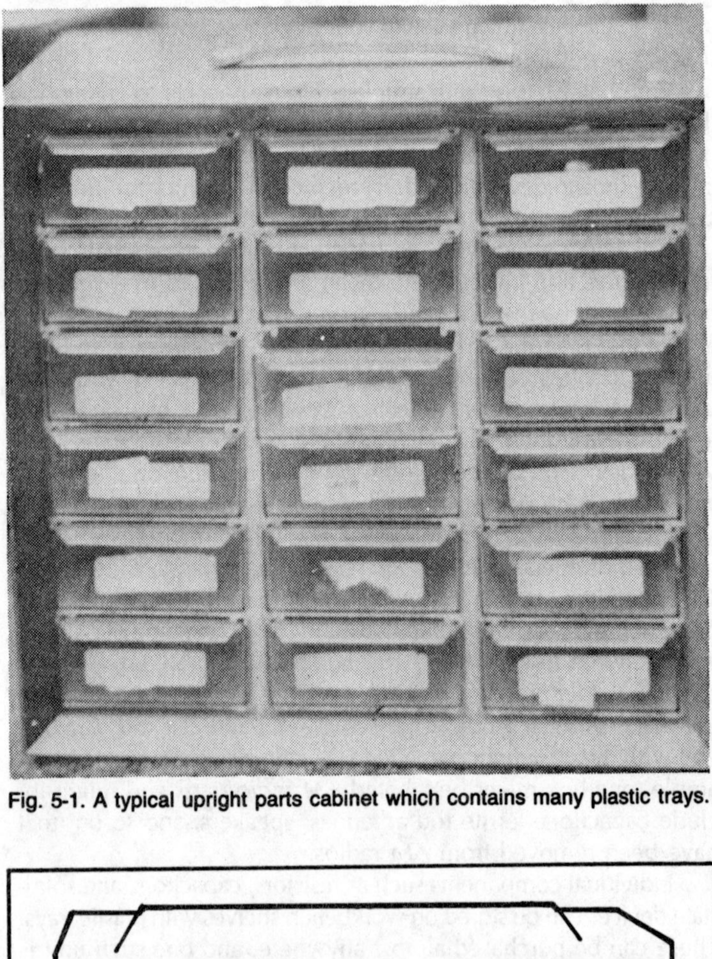

Fig. 5-1. A typical upright parts cabinet which contains many plastic trays.

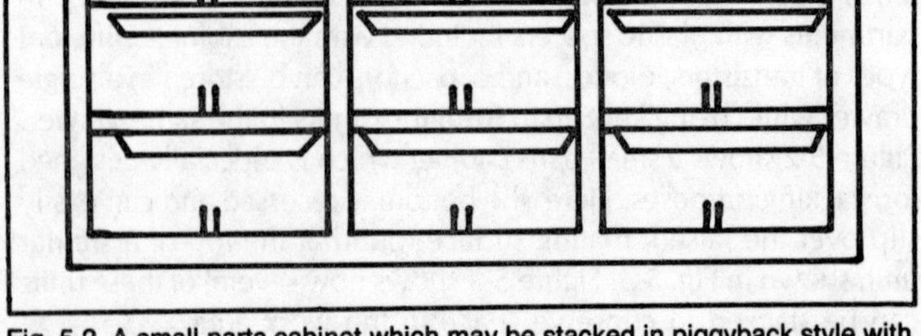

Fig. 5-2. A small parts cabinet which may be stacked in piggyback style with other cabinets.

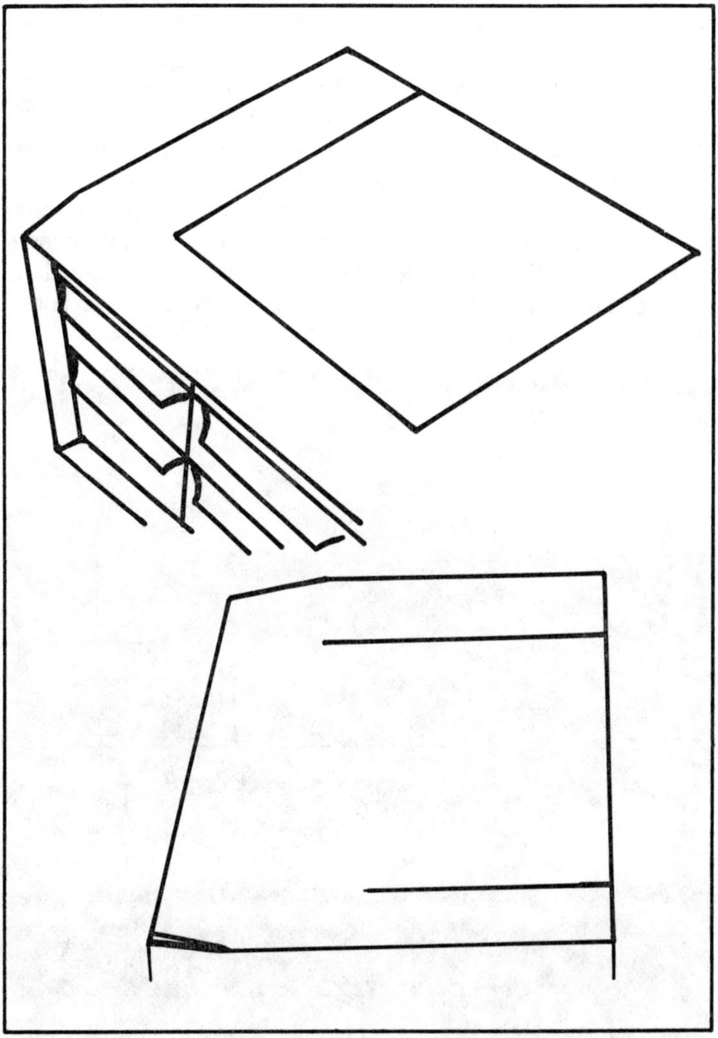

Fig. 5-3. View of the top and bottom of the same parts cabinet showing the recess slots.

ure 5-5 shows a resistor storage cabinet which can be purchased that contains several different compartments in each drawer. Each compartment is marked with a value, which is a great aid to the home experimenter. Ohmic values of carbon resistors are standard for the most part. Since this is the case, resistor cabinets follow the same pattern. If you do not wish to purchase a cabinet specifically for resistors, you can simply made your own and label each drawer as a space for a particular type of component becomes necessary.

Another addition to the electronics workbench that is very useful for storing components is a plastic tray or two. These can be used to house various "pull-out" components which can be properly categorized and stored at a later date. This is where components are placed when originally purchased. When you have the time, go through these trays and store the components in their permanent location. Never let these trays become overcrowded, especially in the case of damaged or useless devices that you know

Fig. 5-4. Piggyback arrangement of two parts cabinets.

Fig. 5-5. A commercial resistor storage tray assembly.

you will never use. Always discard those components that are damaged. These have a tendency to build up and get mixed with useful components. The job of keeping an orderly parts system is primarily comprised of throwing away what you don't need, rather than keeping what you've already filed in a constant state of order. If you do the former, the latter takes care of itself.

If you plan to hold onto any dry cell batteries for any length of time, they should be kept in metal drawers or boxes which are known to be leakproof. Battery storage trays should be periodically checked in order to identify and discard any cells which are leaking. These can quickly damage any good cells remaining, and leakage can ruin other components and finished circuits as well.

The storage system described in these pages should encompass the great majority of parts and components most home experimenters are likely to collect. Of course, there are always some devices and circuit components which don't fit into any category or sub-category and really don't need their own category either. These can be relegated to a miscellaneous box or cabinet. Whenever possible, the materials filed under this category should be placed in a more appropriate area as the situation allows.

With this type of storage system, it should be relatively easy to gather up the parts needed for a specific electronics project in a short period of time. Once you know what you have on hand that can be used, it is a simple matter to order those components you need, and you are then well on your way.

Chapter 6

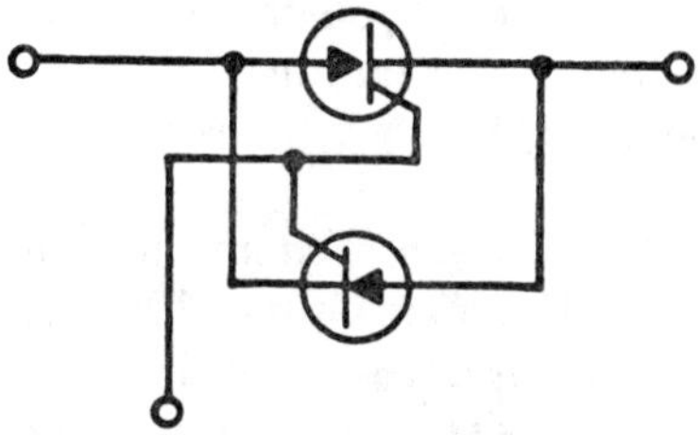

24 Thyristor Projects

THIS CHAPTER INCLUDES THE DETAILED SCHEMATIC DRAWINGS AND circuit discussions you'll need to build electronic circuits using silicon-controlled rectifiers, triacs, and other types of thyristors. Each project is chosen for its usefulness, ease of building, and low construction cost. All of the projects may be built using standard components, many of which are available at your local hobby store. Unusual components are often presented with a suggested manufacturer and a part number. However, these same parts may be readily available through other channels. In most instances, component values may be modified slightly and the device will still yield equally pleasing results.

Many of these circuits involve using potentials derived from the 115-volt ac line. If you have been building a great many solid-state projects using transistors and ICs, there is a good chance that you are accustomed to working with low-voltage circuits and power supplies. When one is faced with a switch to medium-voltage circuits after working with those which present potentials of 25 Vdc or less, there is a potential safety hazard. People who work with low-voltage circuits don't need to be terribly concerned about where they place their hands within the circuit. However, many of the circuits presented here can deliver severe electrical shocks, so extreme caution should be exercised during the checkout and troubleshooting procedures. Don't consider 115 Vac or dc to be relatively harmless. More people are killed each year by this voltage potential than by any other. Many of the circuits in this book have low-voltage sections or legs, followed by those which use the line potential. Be extremely careful for your own personal safety and for that of your test equipment as well. An expensive multimeter can be quickly damaged by applying the probes to a 115-volt potential when the meter scale is set on the 12-volt range.

During troubleshooting procedures, it is best to remove the line plug from the wall completely and discharge any electrolytic capacitors before actually entering the circuit. In some instances, this won't be possible, since checkout may be necessary while the circuit is active. The discussion of voltage measurements in Chapter 3 discusses electronic safety; the practices set out in that chapter should be strictly adhered to.

PROJECT 1: TWO-CHANNEL STEREO LIGHT ORGAN

A light organ or light modulator is a circuit which uses the audio output from a stereo receiver or other such device to control the intensity of an incandescent lamp. This is accomplished by using a silicon-controlled rectifier whose gate circuit is connected to the audio output from the audio device. As the audio level rises, the SCR conducts more and more, and the intensity of the light increases. When the volume falls off, the light dims accordingly. A period of silence at the receiver causes the light to black out completely. A sudden burst of audio will cause an equivalent flash of light. These circuits are often called *light modulators* because light intensities are changed in accordance with the amplitude of the audio output channel.

Figure 6-1 shows the circuit, which is actually two circuits combined. The left channel is represented by the top portion of the schematic drawing, while the right portion is at the bottom. Power is derived from the 115-Vac line, which is wired in parallel with each silicon-controlled rectifier and with X1 and X2, two 115-Vac receptacles. The output voltage to these receptacles will be dc, so only incandescent lamps and other ac/dc devices may be operated. If you connect an ac-only device to either of these receptacles, it will quickly be destroyed.

The input network for each channel consists of a small audio transformer which has an 8-ohm primary and 1000-ohm secondary. These are miniature output transformers which should be available at your local hobby store and can be mounted on a piece of perforated circuit board. The values of the transformer's primary

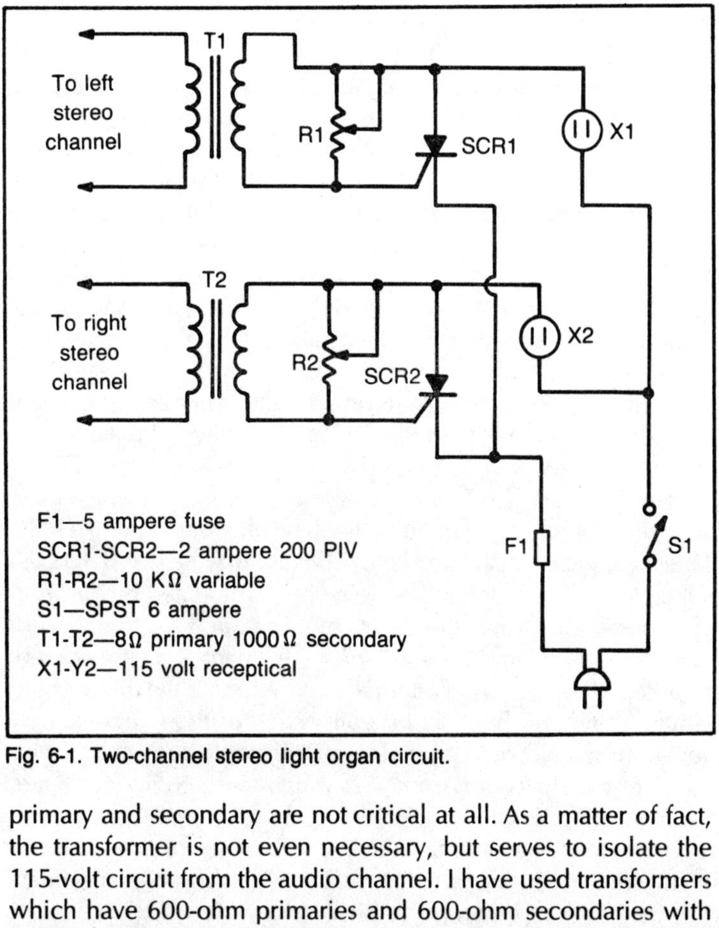

Fig. 6-1. Two-channel stereo light organ circuit.

primary and secondary are not critical at all. As a matter of fact, the transformer is not even necessary, but serves to isolate the 115-volt circuit from the audio channel. I have used transformers which have 600-ohm primaries and 600-ohm secondaries with equally good results, although the values shown will bring about the most efficient operation. Nearly any stereo amplifier has more than adequate drive to control the gate circuits of the silicon-controlled rectifiers.

The total current drain should not exceed about 1.75 ampere at either receptacle. Since two silicon-controlled rectifiers are used, each rated to pass a forward current of 2 amperes maximum, the current drains at each receptacle do not add as far as the SCRs are concerned. In simple terms, this means you may safely draw 1.75 ampere from *each* of the receptacles simultaneously. A 175-watt lamp at each receptacle can certainly be accommodated easily. Some readers might be wondering why the current limit is held to 1.75 ampere instead of 2 amperes, the rating of the silicon-controlled rectifiers. This is done for component protection. The 2-ampere SCR rating is an absolute maximum that must not be exceeded. It's usually a good idea to stay a short distance below maximum ratings to assure long component life. In most instances, you could safely draw a full 2 amperes from each socket, but if this value was exceeded even slightly, the SCRs might be destroyed. If you need more power, replace SCR1 and SCR2 with 3-, 5-, or even 10-ampere units. With these higher values, you will probably find that a lot of heatsinking is required.

Using the components specified in the schematic diagram, the circuit may be built on a 5-inch square section of perforated circuit board. Begin by mounting T1 and T2 near one edge. The secondary winding of each transformer is then connected between the anode and gate electrode of each SCR. Now, clip away any unused portions of circuit board.

The next step is to prepare a mounting enclosure, which may be made of plastic or aluminum. It will be necessary to drill a hole at the back to accept the line cord. The fuse may be wired directly to the circuit board or installed in a chassis-mounted holder. Cut two holes in the back of the case to mount X1 and X2. You'll need to install a two-terminal strip for each channel of the stereo amplifier. These strips are internally connected to the transformer primaries and will be used for connection of the light organ to the 8-ohm speaker output terminals. Drill a hole above each strip to accept the shafts of R1 and R2. When these components are mounted, short lengths of insulated hookup wire are used to connect them to the transformer secondaries. These will serve as sensitivity controls during circuit operation. Also, install S1 in series with the 115-volt input.

When you have completed your circuit, give all wiring a very thorough examination, checking for short circuits and other wiring errors. True, the 115-volt line is fused at 5 amperes, but a wiring error at either SCR may cause these components to be damaged long before the fuse has time to open.

To test the circuit, first connect each input to the left and right channel stereo amplifier outputs. The connections are made in parallel with the leads which normally run to the left and right stereo speakers. With S1 in the open position (off), insert the line plug in the wall outlet. Tune your stereo to a station which is broadcasting music or use a tape player or turntable. Adjust R1 and R2 to their mid-range positions and then activate S1. Adjust the stereo volume control to a comfortable listening position and then adjust R1 and R2 to activate the lights. The lights may be immediately activated when you push S1 on, and it will be necessary to back off on the controls to a point where the music is modulating the lamps. If the lamps remain dark, turn R1 and R2 the other way to increase the resistance. It should take only a few seconds to obtain the proper setting. Variable resistors R1 and R2 are placed across the transformer secondaries and effectively short out the transformers when in the zero resistance position, causing the lamps to glow continuously. As you back off on these controls, however, the audio input begins to control SCR conduction.

If you're faced with the situation of one channel working properly while the other doesn't, remove the plug from the wall and recheck your circuitry in the malfunctioning channel. Also, make certain the balance control on your stereo amplifier is set to the mid-range position so that you have equal output from both the left and the right channel. This is a simple circuit and it should work properly the first time if you've paid close attention to the wiring and have no defective components.

Make sure you build this circuit into an enclosure which does not allow access by unknowledgeable persons. A severe electrical shock will occur if someone touches the SCR leads or the connections to the 115-volt receptacles. Either plastic or aluminum may be used, although I prefer the former due to the ease of mounting the circuit board on this insulating material. If you use aluminum, the circuit board must be mounted using stand-off insulators.

PROJECT 2: LASCR LIGHT METER

A LASCR is a light-activated silicon-controlled rectifier. This means that no external gate circuit is absolutely necessary for triggering, as this can be done with a light source. You can connect a variable resistor between the anode and the gate lead to control the sensitivity the device displays in regard to the incoming light, but in many applications, this is not necessary or desirable. When a small amount of light strikes the sensitized surface of the LASCR, the device begins to conduct. As the light intensity increases more and more, the LASCR's conductivity increases. If the light is decreased, the LASCR's conductivity decreases. Therefore, the output is directly controlled by light intensity.

Figure 6-2 shows a minimum parts circuit for a light meter which uses a common voltmeter, a 9-volt battery and an inexpen-

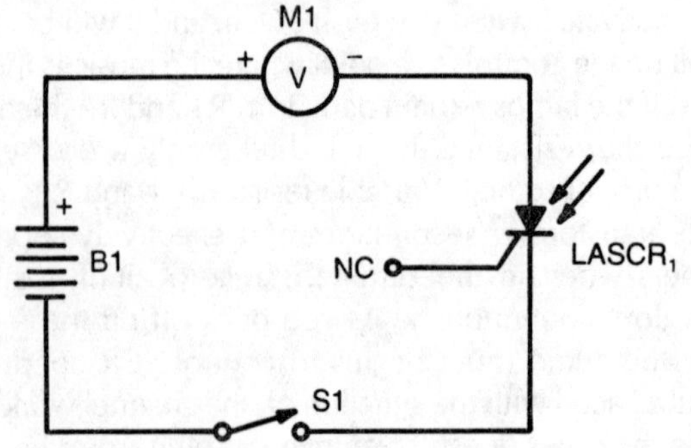

B1—9 volt battery
LASCR$_1$—Radio Shack 276-1095A (MRD920)
M1—0-15 V dc voltmeter (Radio Shack 270-1754)
S1—SPST

Fig. 6-2. LASCR light meter circuit.

sive light activated SCR, which is a relatively new product from Radio Shack. This component is the MRD920, which is designated 276-1095A. It's available for less than $2.00 and is a miniature component which is ideally suited for hobby circuits, including photographic slave strobe units. The device contains three leads for the cathode, anode, and gate and will conduct a peak forward voltage of 200 Vdc. Since this is a small device, its current handling capability is minimal. Current drain must be held to below 300 milliamperes maximum. Due to its 200-volt rating, it can safely be used for controlling circuits that receive power from the 115-Vac line.

In this project, the gate lead (3) is not used at all. You may even clip it away, but the lead can be easily wrapped with tape, as you want to add a variable component to adjust sensitivity later. Circuit power is derived from a 9-volt transistor radio battery, although 22-volt photographic batteries may also be used. The meter has a capability of measuring a maximum of 15 Vdc, so if battery potentials which are higher than this are used, it is necessary to increase the meter range.

In this circuit, the meter serves as the electronic load. When S1 is activated, the circuit is still effectively off, assuming the LASCR is in darkness. However, when a light source strikes the surface of the LASCR, it begins to conduct current. The current flow through the meter causes its indicator to be deflected, which will serve as an indication of light intensity. This type of circuit is best for checking room lighting, and a very bright source will be required to cause the maximum value of 9 volts to show on the meter. The meter scale will be calibrated in volts dc, and you will have to make up a separate chart to extrapolate the intensity readings in light units. This is a basic form of light meter and is best used to indicate relative brilliance. For example, if one light source placed one foot away from the LASCR produces a reading of 6 Vdc on the meter and another light source placed the same distance away produces a 3 Vdc, you would know that the second light source was less brilliant than the first. Sometimes, incandescent light bulbs are used as dummy loads for radio frequency transmitters. With this project optically coupled to the incandescent lamp, a meter indication of output power (relative) is available. There are hundreds of other uses for this circuit, which should cost less than $10. Dramatic savings can be had by using a surplus voltmeter, which may cost only a few dollars.

No circuit board is required for construction of this project. All that's needed is a plastic enclosure with enough panel space to mount the meter and switch and to house the battery. Three tiny holes may be drilled in the top, through which the LASCR leads are inserted. Using insulated hookup wire, make a connection between the negative battery terminal and the anode of the LASCR. Another length of hookup wire connects the cathode to one contact of S1. The negative terminal of the 9-volt battery is connected to the remaining switch contact and the positive battery terminal is connected to the positive meter terminal. To use the device, focus the light on the surface of the LASCR. You can use a small cardboard tube which is fitted around the LASCR and glued to the case. A piece of electrical tape may be placed over the top of this to shield it from all light when the circuit is not in use. With the tape removed, a beam of light may be directed down the tube to strike the LASCR surface. To conserve battery life, always flip S1 to the off position when the circuit is not in use. Due to the value of B1, the meter indicator will never be driven to full scale. Even better sensitivity can be obtained, however, by locating another meter type which offers a 0-10 volt scale.

A fair amount of light is needed to get the LASCR to conduct fully. Better sensitivity can be achieved by connecting a 100 K variable resistor between the negative terminal of M1 and the gate lead. This can be adjusted to bring about the desired triggering range with a reference light source directed on the LASCR.

PROJECT 3: SCR TIMER CIRCUIT

A timer circuit can be made using many different electronic components by depending upon the discharge of an electrolytic capacitor through a resistor. A capacitor has the capability of storing electrical energy for long periods of time, and if passed through a resistor in parallel with its positive and negative leads, the stored energy will be bled off gradually at a rate that can be fixed and measured. The actual rate of discharge will depend upon the value of the capacitor and that of the resistor.

The circuit shown in Fig. 6-3 uses a silicon-controlled rectifier

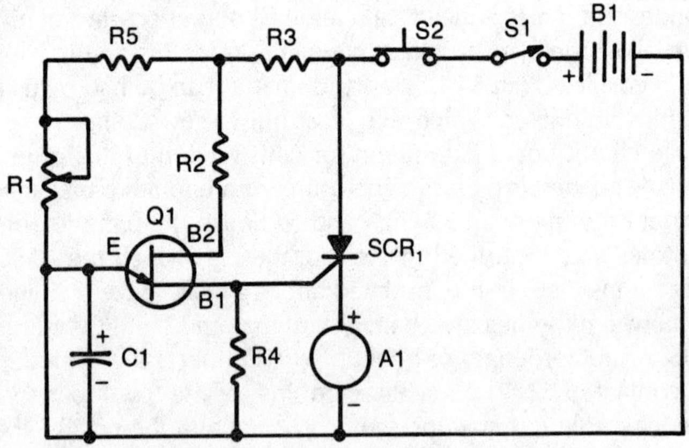

A1—Piezoelectric buzzer (Radio Shack 273-065)
B1—9 volt battery
C1—480 μF electrolytic 25 Vdc
R1—1 megohm variable
R2—200 Ω 1/2 watt
R3—620 Ω 1/2 watt
R4—120 Ω 1/2 watt
R5—100 K Ω 1/2 watt
S1—SPST
S2—*Normally closed* momentary switch
SCR$_1$—50 PIV 1 ampere SCR
Q1—ECG 6400

Fig. 6-3. SCR timer circuit and component list.

whose conduction is controlled by a unijunction transistor. This transistor fires when C1 has been discharged to a certain value. The discharge path is through R1, a variable resistor, which can be adjusted to deliver different discharge rates and thus, differing triggering times. When S1 is initially activated, current flows from the 9-volt battery. In an extremely short period of time, C1 will be fully charged and will then begin to bleed off through the resistance. When the capacitor is sufficiently discharged, Q1 conducts and activates the silicon-controlled rectifier. This causes current to flow through a sonic transducer (A1), which will sound. The timer can be reset by depressing S2, which is a normally closed momentary switch. R1 is adjusted to deliver the desired timing rate. Using the circuit components shown, you should be able to obtain a maximum time lag of about eight to ten minutes.

I used a Radio Shack piezoelectric buzzer as an indication that the timing period had elapsed. Radio Shack offers several different types of buzzers. The one I chose was designed to be mounted on perfboard and draws only about 10 milliamperes at 9 volts. Many other buzzers and audio transducers may be used here, as long as they can be powered from a 9-volt source.

The circuit is constructed on perfboard, and I suggest starting with a 6-inch square section. Mount all circuit components as shown and make sure you correctly identify the leads of the unijunction transistor, Q1. The silicon-controlled rectifier can be any type. The unit specified has a forward current of 1 ampere, which is common, but much smaller values may also be used, since the circuit draws little current when the buzzer is activated. Make certain the buzzer is connected with its positive lead at the cathode of SCR1 and its negative lead at the negative battery terminal or circuit ground. Wiring is relatively non-critical. Be polarity-conscious throughout the process, however, as B1, Q1, SCR1, and A1 must be connected with the polarities as shown in the schematic diagram. The same applies to C1, which is a polarized electrolytic capacitor.

To test the circuit, adjust R1 to the mid-range position and activate S1. Slowly decrease the resistance of R1. After a short time, the buzzer should sound. Then increase the resistance of R1 slightly. The buzzer should go off after a longer period of time. If you get no response, make sure your battery is good and reexamine all wiring for an error, especially in regard to the polarities of components. To reset the alarm, simply press S2, and the timing sequence begins all over again. Using the circuit as shown, it is necessary to fit R1 with a pointer knob and a scale which can be calibrated in minutes, seconds, or whatever. It is quite impossible, even with the knob, to reset R1 exactly to a previous point which yielded a certain time interval. Therefore, this timer circuit cannot be used to accurately time events or sequences to within a second or so. Rather, it is intended as an indication of approximate time intervals. Now, if you had a more exacting use for this circuit in mind, don't be alarmed. With a few modifications, the circuit can yield highly accurate fixed time intervals. Of course, if you want to set the present device up to time out exactly five minutes or one minute, this is no problem. Just adjust R1 until this time interval is obtained. However, when you want several different time intervals, the fallacy of this circuit shows itself.

To obtain an accurate multi-interval timer, simply include more variable resistors, which can be switched into the circuit with a rotary switch. Alternately, you can initially determine the resistance needed by measuring the value of R1 with an ohmmeter and recording the resistances on a sheet of paper. You may then replace R1 with a number of fixed resistors of the needed values. Even here, it may be necessary to install trimmer resistors in series with the fixed ones to allow for precise adjustment. For even more accurate timing, replace C1 with a parallel combination of high-quality nonpolarized capacitors. These can be purchased inexpensively from Radio Shack, but many of them will be required to obtain the same size as C1. Alternately, you can simply increase the size of R1 to several megohms and use a smaller-value capacitor to bring about the same timing intervals.

PROJECT 4: SCR TRANSMIT INDICATOR

A transmit indicator is a circuit, independent of the radio frequency transmitter, which is used to determine the presence of rf output. These are often used in setups where the transmitter is remotely located from the actual operating position. The circuit is not directly connected to the transmitter in any way, but is actually a miniature radio receiver of sorts which detects the rf output and causes the triggering function to result.

The circuit is shown in Fig. 6-4. If you're familiar with crystal radios, you'll immediately recognize the detector input circuit. The antenna samples a portion of the rf energy, and the germanium diode (D1) rectifies this, charging C2. The dc output from the detector is used to trigger a small silicon-controlled rectifier, which allows current to flow through a 6-volt flashlight bulb. This serves

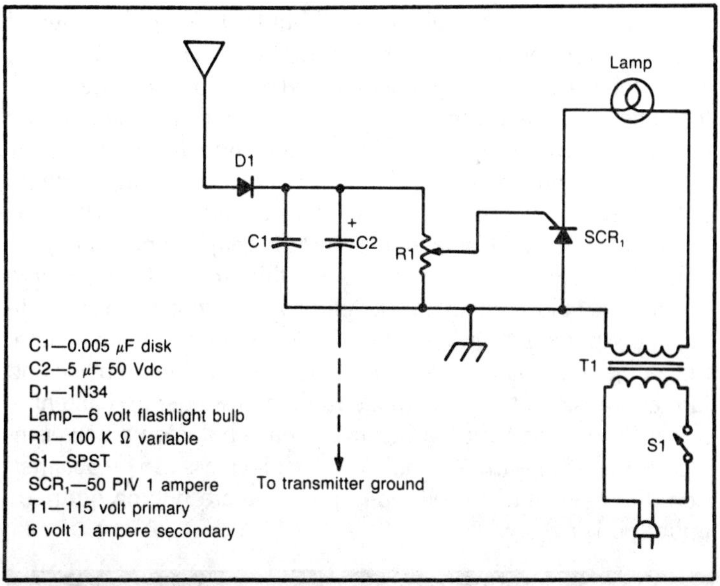

Fig. 6-4. SCR transmit indicator circuit.

as an indication that the transmitter is indeed putting out power. In low-power applications, such as those associated with citizens band radio, it will be necessary to couple the antenna very closely to the CB antenna. Power is supplied to the lamp by a 6.3-volt transformer, which is driven by the ac line. A buzzer can also be used to place of the lamp, and by removing C2, an audible transducer may be included to serve as a sidetone generator. This can be used with transmitters to monitor code transmissions. Each time the transmitter is keyed, so is the transducer or buzzer, and its duration corresponds to the dot or dash produced at the code key. You must remove C2 from the circuit completely, however, because in its present form, this device is set up to act only as an indication that power is being radiated from the transmitter. Due to the charging effect of C2, the light will stay on for a short while after all transmissions have stopped.

You can build this circuit on perfboard and even mount the transformer here, although you'll probably want to install the transformer and its primary circuitry to the bottom of the enclosure which is used to house the remainder of the circuit. Adjustable resistor R1 serves as a sensitivity control. For amateur radio uses, C2 can probably be discarded altogether; it is there to provide gate current in low-power applications.

Wiring is not especially critical, but try to keep the detector circuit portion separate from the portion which controls the lamp. The antenna can be a length of insulated hookup wire. (The longer the better.) You may also use a telescoping whip antenna salvaged from a defective CB walkie-talkie. The operation of this circuit may depend to a large extent upon the ground connection, although different setups will bring about different triggering. This is the type of circuit you build and then adjust in conjunction with your transmitter. Through precise use of R1, this circuit can even serve as a remote field-strength meter. Adjust your antenna and transmitter, and while transmitting, adjust R1 to a point just below where triggering of the lamp occurs. The circuit is then left in place and further adjustments can be made to the transmitter and antenna. If these cause the light to trigger, you have obtained a power increase. Keep adjusting R1 until no further increase can be obtained from the transmitter. Then, you know you are putting out maximum power.

PROJECT 5: SCR BURGLAR ALARM

Burglar or intrusion alarms are being used more and more today, not only in businesses, but in private homes as well. These devices are designed to sound an alarm when a door or window is entered. Some are extremely expensive, while others are quite cheap. The latter tend to be ineffective in many instances, but can serve as excellent backup alarms to a larger system.

Figure 6-5 shows a simple burglar alarm circuit that uses the latching effect of a triggered SCR when used with a dc supply to deliver a quite effective intrusion alarm. The circuit uses a Sonalert transducer to produce a loud tone when triggered.

The circuit is straightforward and can be assembled in a short period of time with a minimum of components. It is self-contained and normally draws no current when in the armed state. S1 is the control switch which arms the circuit, while S2 is an alarm switch that is normally open when the door or window is secure. This is often a magnetic reed switch which is held open by a separate magnet attached to the door or window. When entry is attempted, the magnet is moved by raising the window or opening the door, and the switch snaps shut. This supplies current to the gate circuit of SCR1, and current is allowed to flow through A1. The SCR latches to the on position and will remain on even if S2 is opened again. The only way to disarm the circuit is to remove all power by switching S1 to the off position. A 6-volt lantern battery is used to power the circuit and should provide many hundreds of hours of operation, since the circuit will be triggered only when an intrusion occurs.

You can build the circuit in a single enclosure and then use hookup wiring to connect the remotely-located alarm switch. Alternately, the entire device can be installed directly at a door or window or other such location where intrusion could occur.

To test the finished circuit, make sure S2 is in the armed state or open, set R1 to a mid-range position, and activate S1. If the alarm goes off immediately, this probably means that S2 is closed or that there is a wiring error. Now, close S2 and the alarm should sound. If not, readjust R1 for a lower resistance. When the alarm goes off, continue adjusting R1 to a slightly lower resistance position and then switch S1 to the off position. The alarm will be turned off and you can now rearm the circuit by opening S2 again and activating S1. When S1 is flipped on, the alarm should not sound.

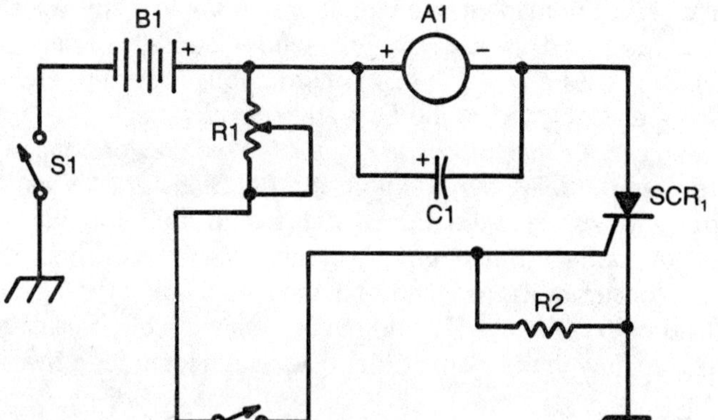

A1—Sunalert (Allied Elect #854-6545)
B1—6 volt lantern battery
C1—100 μF 16 Vdc
R1—10 K Ω variable
R2—1 K Ω 1/2 watt carbon
S1—SPST
S2—Alarm switch (open when armed)
SCR_1—5 K 3950 (RCA)

Fig. 6-5. SCR burglar alarm circuit.

When you close S2 by opening the door or window, the alarm will sound.

While this circuit can be used to protect only one possible entrance to your home, several other entrances can also be protected by wiring other normally open switches in parallel with S2. If any one of them closes, the circuit will be activated.

PROJECT 6: ANOTHER SCR ALARM

The previous alarm circuit used normally open switches for activation. While these are useful in simple circuits, they do present somewhat of a logistical problem, in that a continuous loop of alarm switches cannot be used. The loop switch arrangement uses several normally closed switches in series, and when any one of them is opened by an intruder, the circuit is broken and the alarm is tripped. Using the previous circuit, it was necessary to run a pair of wires directly from the alarm portion to the area to be protected.

The circuit shown in Fig. 6-6 avoids this necessity, as only two connections are made to the circuit proper for as many switches as you desire. This can allow a whole house, upstairs and downstairs, to be protected by a continuous loop, the ends of which are connected to the SCR and power supply.

You can see that the circuit is similar to the previous one, with the exception being the addition of a pnp transistor. When the alarm switches are in the normally closed position, the transistor does not conduct. But when one of the switches is opened, conduction occurs and current flows in the circuit branch at the emitter lead of Q1, through R2 and the SCR gate circuit. This causes current to flow in the main SCR circuit, which includes a Sonalert alarm.

Sensitivity is controlled by adjusting R1. Unlike the previous circuit, this variable resistor controls the switching sensitivity of Q1. R2 is a fixed value resistor that will cause SCR1 to conduct when current flows through this device.

The circuit is built much like the previous one, using perfboard and mounting everything but the alarm switches in a single enclosure. A lead is then brought from the base of Q1 to the first alarm switch. The remaining alarm switch terminal is connected to the next alarm switch, and that to the next, and so on. The final alarm switch contact is connected to the anode of SCR1. When all possible entrances to the home are secure, the alarm switch contacts are closed and the transistor does not conduct. Current is being conducted through the closed loop switches. When any of the switches are opened due to an intrusion, this portion of the circuit is broken and current flows through Q1 and thus, to the gate circuit of SCR1, activating the alarm.

Once the circuit is complete, make sure all alarm switches are closed, set R1 to the mid-range position, and activate S1. If the alarm sounds immediately, increase the value of R1 slightly after deactivating S1. You want to arrive at a resistance value which will not turn the alarm on when S1 is activated and all alarm switches are closed. When you have reached this point, open one of the alarm switches and the alarm should sound immediately. If not, back off a bit on R1, but not to a point where the alarm is prematurely triggered.

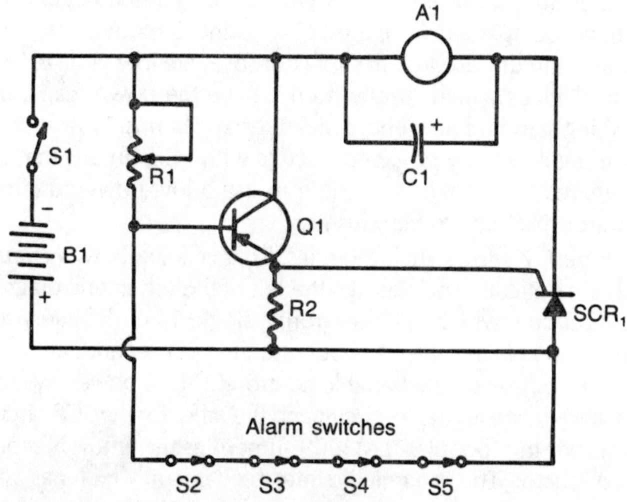

A1—Sonalert (Allied Elect # 854-6545)
B1—6 volt lantern battery
C1—120 μF 16 V electrolytic
R1—1 megohm variable
R2—1000 Ω 1/2 watt carbon
S1—SPST
S2-S5—normally closed alarm switches
Q1—SK 3466 (RCA)
SCR$_1$—SK 3950 (RCA)

Fig. 6-6. SCR burglar alarm circuit using a closed detection loop and component list.

This circuit is far more practical when it is necessary to protect a number of entrances to a home or business, but current will flow while the device is in the armed state. Therefore, it's a good idea to use a 6-volt rechargeable battery for B1 and connect it to a charger which is continuously activated. You could use a 6 Vdc power supply driven by the ac line, but in the event of a power failure, your home would not be protected. Using a battery and a charger, you are assured of a full charge at all times, and in the event of a power failure, there is enough stored energy in the battery for many days of operation.

PROJECT 7: INCANDESCENT LIGHT FLASHER

Many integrated circuits can be used to periodically flash light-emitting diodes and small flashlight bulbs. Although these may serve a useful purpose, you may want to flash much larger bulbs which can be operated from the 115-Vac line. Unfortunately, most integrated circuits used for flashing purposes are low-voltage, low-current devices which simply do not have the power capability of flashing standard household incandescent lamps. However, by combining these low-powered circuits with SCR circuits that can be triggered by light, it is possible to use a low-powered circuit to control the high-powered one.

Figure 6-7 shows such a circuit. This is actually two circuits which are optically coupled. To the left of the schematic diagram is the IC circuit, which operates from a single 1.5-volt battery and will flash an LED at a rate of about one time per second, although this can be adjusted with variable control R3. Two power switches are included, one for the IC circuit and the other for the SCR circuit.

I used the popular LM3909 timer/flasher from National Semiconductor. This is a reliable integrated circuit which has been used in a large number of flasher projects. An LED is coupled to the output of the IC, which completes this portion of the circuit.

To obtain high-powered flashing capability, the glow of the LED is closely coupled to a small photocell, available from Radio Shack for a few dollars, installed in the gate circuit of SCR1. The photocell is more appropriately called a *cadmium-sulfide cell*, which is a light-variable resistor. When the sensitized surface of the photocell is in total darkness, the device is equivalent to a 500,000-ohm resistor. However, as the light level is increased, the internal resistance decreases, allowing more and more current to flow in the SCR gate circuit. A variable resistor is also included in series with the photocell and the SCR gate to serve as a manually

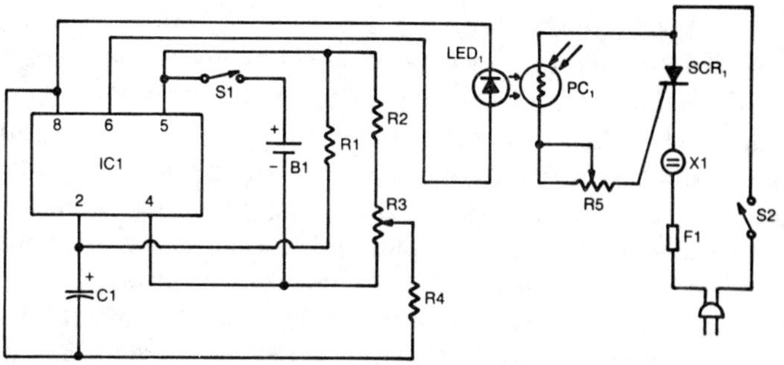

B1—1.5 volt D cell
C1—420 μF electrolytic 10 Vdc
F1—5 ampere fuse
LED$_1$—yellow LED
PC$_1$—Photocell (Radio shack 276-116)
R1 + R4 68 Ω 1/2 watt carbon
R2—2700 Ω 1/2 watt carbon
R3—2500 Ω 1/2 watt variable
R5—10 K Ω variable
S1—SPST
S2—5 ampere SPST
SCR$_1$—5 K 3570 (200 PIV - 4 amperes)
X1—115 volt receptical
IC1—LM 3909 (National Semiconductor)

Fig. 6-7. Incandescent light flasher circuit and component list.

adjustable sensitivity control. When the value of PC1 is brought low by the flash of the light-emitting diode, the SCR fires and conducts direct current to the 115-volt receptacle.

A 4-ampere silicon controlled rectifier has been used to allow you to plug in lights which total a maximum of 375 watts. You can use a lower-powered SCR if you wish to control low-wattage lamps. The 4-ampere SCR specified must be mounted to a heatsink, and the appropriate insulating materials are usually included with the device. Coat the SCR and the thin dielectric sheet with silicone grease before bolting the device to the heatsink. Radio Shack makes several sinks that can be attached directly to perfboard.

Begin construction by mounting the IC on a 5-inch square section of perfboard and connecting the various components associated with this portion of the circuit. When the IC circuit is complete, activate S1 and adjust R3 while observing the LED. You should be able to bring about a flash rate which is very slow or one which is so fast that the LED seems to glow almost continuously. If you cannot get the LED to flash, then you've probably made a wiring error or have a defective component. This circuit portion must be operational before moving on to the SCR circuit.

SCR1 is also installed on the circuit board, mounted to an appropriate heatsink. Connect the photocell between the anode of the SCR and the adjustable resistor. Place its sensitized surface right up against the LED. This portion of the circuit is completed by wiring in the receptacle, the fuse block, S2, and the terminated line cord. Before inserting the line plug, temporarily short circuit PC1 with an alligator clip lead and adjust R5 to its maximum resistance position. Plug in an appropriate light at X1, insert the line plug in the wall, and activate S2. Slowly decrease the value of R5 until the light comes on. This assures you that the SCR portion of the circuit is operational.

Now, remove the jumper at PC1, reinsert the line plug, and activate S1 and S2. Make certain the LED is flashing properly and adjust R5, if necessary, to bring about the activation of the incandescent lamp, which is attached at X1. If you have problems getting the incandescent lamp to flash, even with R5 set to a minimum resistance position, try installing a 10-kohm half-watt resistor in parallel with PC1. This will lower the overall gate circuit's resistance and may bring about better triggering sensitivity.

The completed circuit may be used for special lighting effects, especially in the entertainment field. The circuit may even be coupled to one of the previous intrusion alarm projects to serve as a visual indication of an intrusion or to possibly scare the intruder off.

PROJECT 8: RECEIVER SIGNAL ALARM

Shortwave listeners often spend hours on end monitoring a specific frequency for a particular radio station. These stations may transmit irregularly, so it is necessary to be constantly on guard. Many shortwave listeners can recall sitting in front of their receiver with headphones clamped tightly on, passing away many hours without ever hearing the much-sought-after station. Then too, many foreign stations may transmit at regular hours, but band conditions may be such that the signal is readable only when atmospheric conditions are favorable.

If there were some way you could monitor a certain frequency without actually having to be right there on the scene, shortwave listening might become a little more practical. The circuit shown in Fig. 6-8 may take some of the strain off the shortwave listener, although it does have certain limitations. This is an SCR-triggered alarm which sounds when audio is present. The circuit is designed to be plugged into the receiver headphone jack, and a sensitivity control is provided to prevent triggering from atmospheric static. Thunderstorms, with their high static electricity content, may trigger the alarm anyway, but you need to run to the receiver only when the alarm comes on and stays on, indicating a continuous audio presence.

Looking at the circuit, you can see the audio output from the receiver is connected to the primary of T1. The primary should match the output impedance of the headphone jack and can be varied to suit an individual receiver. The energy is fed from the transformer secondary to the SCR gate circuit, which, when triggered, supplies power to X1, a 115-volt receptacle. Here, an electric lamp (or some other type of dc device) can be attached.

To use it, attach the circuit and tune the receiver to the appropriate frequency. R1 is set just below the point where the indicating device is triggered. Then, when your receiver picks up the audio from the transmitting station, the alarm is triggered, and you can listen to your heart's content. You can install a DPDT switch at the transformer secondary, which allows you to switch from the monitor circuit to the headphones. Alternately, you can simply remove the monitor jack and insert the headphone jack in the receptacle.

You can also use this monitor in a different way by connecting a dc latching relay to the output receptacle. The relay should be of the normally closed variety. A communications receiver will be required which contains a bfo. This will allow you to tune in

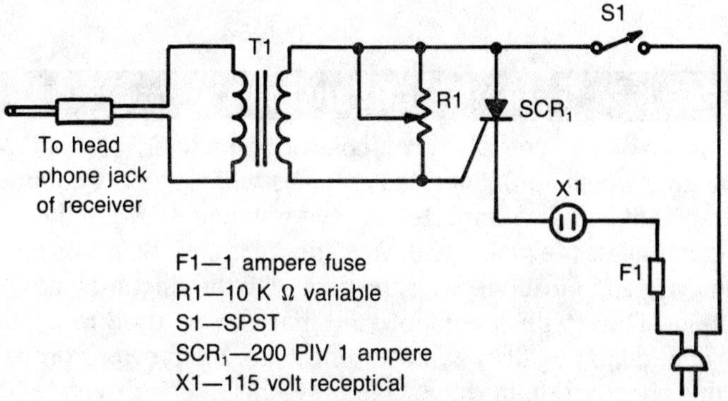

F1—1 ampere fuse
R1—10 K Ω variable
S1—SPST
SCR$_1$—200 PIV 1 ampere
X1—115 volt receptical
T1—500 primary-500 Ω secondary

Fig. 6-8. Receiver signal alarm circuit.

the carrier of an AM broadcast station which will produce a solid tone at the receiver output. This circuit then becomes an off-the-air monitor which can be used anywhere within the coverage area of the AM station. The tone causes the SCR to trigger and feeds power to the normally closed dc relay. When a normally closed relay is activated, its contacts open up. A light or other alarm should be connected across these contacts. Therefore, as long as the transmitting station is on the air, the relay is activated, opening the contacts to the alarm. However, if the station goes off the air for even a few seconds, the SCR stops conducting and the normally closed relay springs shut. This closes the alarm circuit contacts, making it go off. The relay can control a buzzer or some other audible device which could be heard from any room in the home. Chief engineers at commercial broadcast stations often use devices like this when they are on call.

This circuit cannot be used in this manner with FM broadcast stations or with single sideband transmissions which suppress the carrier. The monitor can be altered to serve as a "loss of audio" alarm by installing a 50 P.I.V. diode and a 50-microfarad filter capacitor at the transformer secondary. As long as audio is present, the SCR circuit remains in the conducting state. When audio is lost for a period of time which is greater than the discharge rate of the filter capacitor, the alarm sounds. A 50-microfarad capacitor may not be large enough to prevent false triggering due to pauses of a few seconds which are standard in the commercial broadcasting field. A larger value capacitor allows for audio pauses of longer durations without falsely triggering the relay attached to X1.

PROJECT 9: TRIAC LIGHT DIMMER

A triac is a type of silicon-controlled rectifier, or more accurately, two silicon-controlled rectifiers which share a common gate circuit and are connected in reverse parallel. When the gate circuit triggers the triac, the device conducts current in two directions instead of just one, as is the case with the silicon-controlled rectifier. Due to this operation, the triac can be used to control the conduction of alternating current. All of the previous projects in this chapter output dc voltage only; this one is driven by the 115-volt line and will output a maximum value of 115 Vac, but is adjustable by means of the variable resistor to deliver much lower values. This makes the circuit ideal for motor speed control ap-

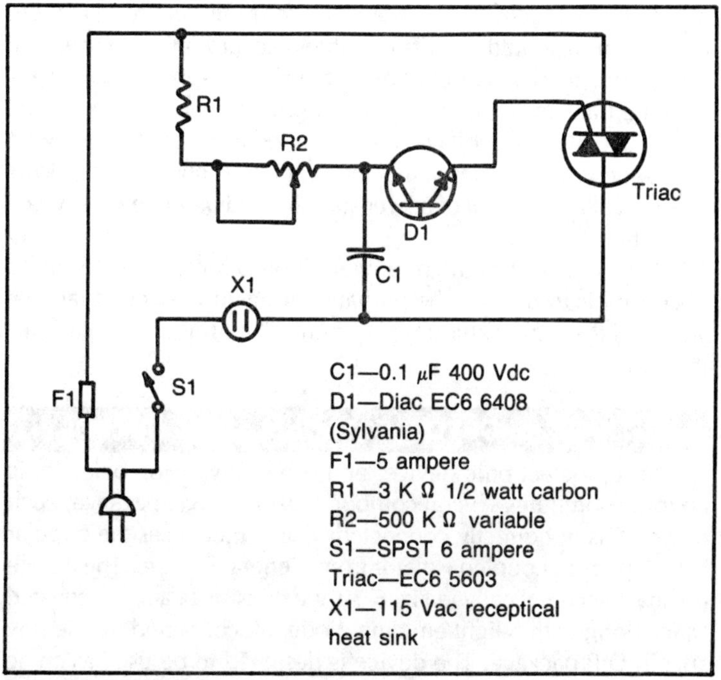

Fig. 6-9. Triac light dimmer circuit.

plications, dimming lights, and controlling other alternating current devices.

The circuit is shown in Fig. 6-9 and is quite similar to many of the SCR circuits presented earlier. Indeed, in some applications, the triac may be used as a replacement for an SCR when an ac output is desirable. This circuit uses a device called a *diac* to provide triggering of the SCR gate circuit. This brings about positive triggering control. The actual value delivered to the 115-volt receptacle (X1) is determined by the setting of R2. R1 is a protective resistor which assures that the full 115-volt potential is not delivered to the diac and gate circuit. C1 is the trigger capacitor which makes the SCR fire when its value rises to the breakover voltage of the diac. When the triac fires, C1 is discharged, but then recharges again through R1 and R2 until the next firing sequence. The triac used is a 4-ampere type that requires external heatsinking. The entire circuit can be built on perfboard and mounted in an aluminum enclosure which can serve as a heatsink (with proper electrical insulation) and for mounting R2, S1, and X1.

To test the circuit, connect an electric lamp to the built-in receptacle and activate S1. Adjust R2 for minimum to maximum and the lamp should alter its brightness accordingly. You can use this circuit for voltage control to transformer-operated devices. When a transformer with a 115-volt primary is connected to the receptacle, its drive voltage may be varied from low to high, with corresponding variations in the output at the secondary. This allows precise voltage control of power supplies and other devices which use a transformer.

This circuit has high power applications, and all that's necessary is to beef up the primary wiring and select a triac that will pass the higher amount of current. The firing circuit can stay the same.

PROJECT 10: OPTOELECTRONIC SCR CONTROL

An optoelectronic device, and especially an optoelectronic coupler, combines several components in a single package, each of which is not directly connected. This project uses the popular H11J1 photon-coupled isolator from General Electric. The device is shown schematically in Fig. 6-10 and consists of a light-activated triac, along with a light-emitting diode, all contained in one tiny six-pin DIP package. The device is designed to be used as an ac switch which is dc-controlled and offers tremendous electrical isolation between the two circuit portions. This particular isolator provides about 4000 volts isolation between the light-emitting diode and the triac.

The circuit using this device is shown in Fig. 6-11 and is ex-

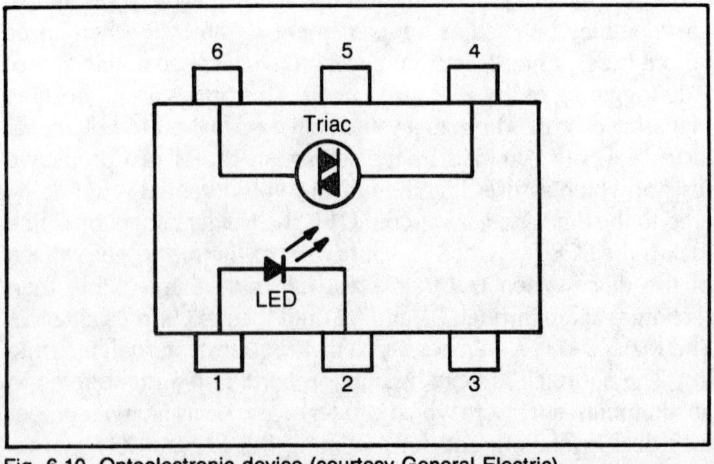

Fig. 6-10. Optoelectronic device (courtesy General Electric).

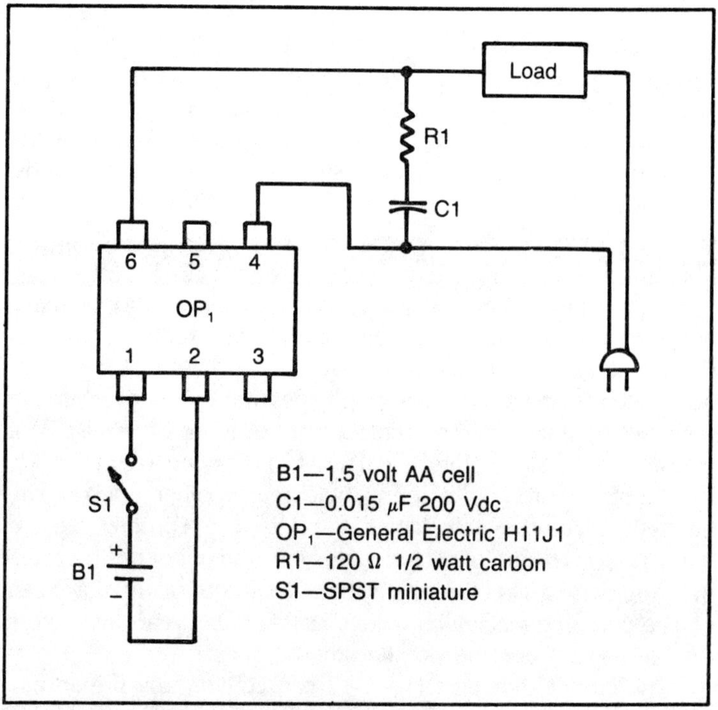

Fig. 6-11. Optoelectronic SCR control circuit.

tremely tiny owing to the small size of the General Electric device. It can easily be installed on a 2-inch square section of perforated circuit board. The switching portion of the circuit is found at terminals 1 and 2. Here, a 1.5-volt penlite cell is connected in series with the two terminals and S1 controls the power to the internal light-emitting diode. The 115-Vac portion of the circuit is found at pins 4 and 6. Two other components, R1 and C1, are included for transient protection, but are not absolutely mandatory. Using this circuit, you can switch on an ac device with the 1.5-volt circuit. Remember, there is no direct connection between the switching circuit and the 115-volt circuit.

Unfortunately, the tiny optoelectronic isolator is a low-power device and you will have to limit the 115-volt load to less than 75 milliamperes. However, you can use a 115-Vac relay which draws less than this amount to control large amounts of power. This type of circuit is safe for relay switching high-voltage circuits up to a maximum of about 4,000 Vac or dc. The protection here is within the isolator, which protects the low-voltage switching circuit from these high potential values. You may also use this circuit to control a separate SCR or triac which is able to pass a much higher amount of current. Optoelectronic isolators are in common usage today and serve as an easy means of switching on high-voltage circuits without having to resort to large and expensive mechanical or electromechanical switches that offer the same degree of electrical isolation.

PROJECT 11: SCR GARAGE DOOR OPENER

Many persons have installed motorized garage door openers, and the more expensive versions of these often offer remote activation control through small radio transmitters. Others are automatically triggered when the lights from the automobile are detected by a sensor. The remote control types of garage door openers are certainly quite useful and avoid having to exit the protection of the automobile during inclement weather to trip a manual switch. Unfortunately, their high price is often prohibitive, and many persons resort to installing the type which requires the manual tripping of a switch for activation. Then too, many motorized garage door openers which were installed many years ago before remote control became popular are still in use.

The circuit shown in Fig. 6-12 can convert any present garage door opener to remote controlled switching by combining it with an existing system. Using this circuit, the headlights from your automobile will trigger the motor, and once your car is inside, the door can be lowered by flipping the switch manually.

The circuit uses the inexpensive Radio Shack LASCR, which was used in a previous project. The finished circuit will be extremely small and is self-contained, so it can be mounted as a discrete unit on the front of the garage near the door opening. An internal 9-volt battery supplies power, and a momentary switch is provided to manually lower the door or raise it, should a light source not be available.

The LASCR is connected to the circuit using only its anode and cathode leads. The gate lead is not used at all, as a source of light will be used to cause the device to break into conduction. When this occurs, current flows through a circuit board-mounted relay, which closes its contacts. These are located in series with a 115-Vac latching relay, which is available from Allied Electronics. This particular relay latches closed with the first burst of ac current and opens again upon receiving another burst. The contacts of the second relay are connected in series with the power line to the motor. The second relay is necessary because of the current most motors draw during operation and because it is a latching type, which is necessary for positive triggering. When the circuit board relay closes due to the glow of the headlights, the second relay latches firmly into place, triggering the motor. If a latching relay is not used, the movement of the automobile, and thus, its lights, might cause the LASCR to revert to the off state and leave the door in a partially opened position. The circuit board relay is not a latching type and will be dependent upon the headlights to keep it closed. In this circuit, however, a closure of only a second or so is required to initially trigger the latching relay, which will remain in the on state. Most garage door openers are equipped with automatic cutoff switches which are activated when the door is fully opened.

After your vehicle is safely in the garage, press the manual close switch and S2 at the LASCR circuit to reset the light detector. In some cases, it will be necessary to depress S2 first, since

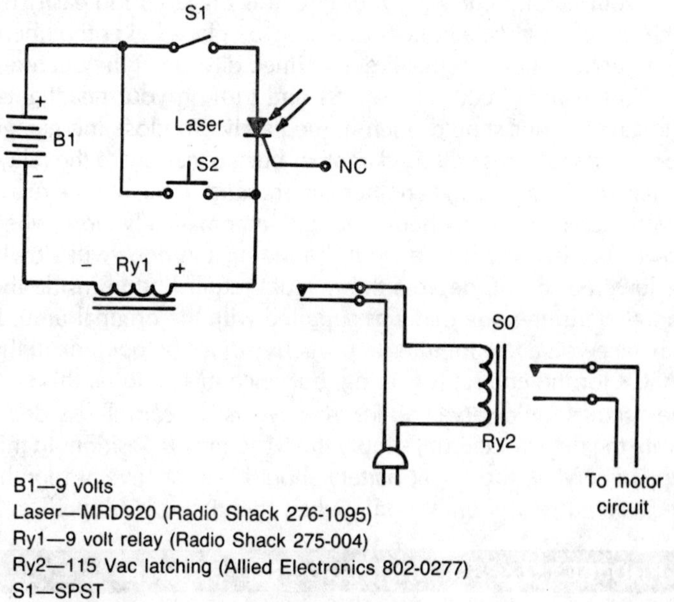

B1—9 volts
Laser—MRD920 (Radio Shack 276-1095)
Ry1—9 volt relay (Radio Shack 275-004)
Ry2—115 Vac latching (Allied Electronics 802-0277)
S1—SPST
S2—Momentary switch (normally open)

Fig. 6-12. SCR garage door opener circuit.

the latching relay is still locked in the on position. Individual systems may require different connections from the latching relay, but this is usually quite simple, since most openers allow provisions for alternate control circuits.

The detector circuit may be built on a 4-inch square section of perfboard. This will be used mainly for mounting the relay (RY1) and the battery. The circuit board can then be enclosed in an appropriate container, which should be shielded from the elements. You'll need to drill a hole in the case to access the opening of the LASCR. A piece of clear plastic material can be glued to the inside of this opening to make the assembly watertight and still allow the entrance of light. Since it is necessary to expose the case of the LASCR in order for it to be struck by a beam of light, this device should be mounted close to the surface of the enclosure.

By using the LASCR in its least sensitive operational mode, false triggering of the circuit is usually completely avoided during daylight hours. It takes a fairly intense beam from the automobile's headlights to effectively trigger the circuit, and it is necessary to adjust the window opening of the enclosure for proper alignment with your automobile(s). If your circuit is triggered too easily by other nearby lights, a hollow tube may be placed over the opening to deflect all lights that are not aimed directly at the opening.

To test the circuit, activate S1 and turn on your headlights. The garage door should open immediately. To close the garage door, press S2. This will send another burst of current to the relay, which will, in turn, send another current burst to the remote relay. If your garage door opener is set up to automatically close when power has been removed from the motor, the door will slowly be lowered. If not, depress the manual return button inside the garage door (the one that was supplied with the original unit). It may take you a few minutes to properly adjust the opening to the LASCR for the correct triggering, but once this is done, this simple circuit should serve you for many years. To completely deactivate the detector circuit, simply flip S1 to the off position. In this type of service, the 9-volt battery should last for several months because current is drawn only when the circuit is tripped.

PROJECT 12: SOLAR-TRIGGERED SWITCH

Many of the previous SCR and triac projects have used a fixed or variable resistor between the anode and the gate of each device in order to bring about triggering or to set trigger sensitivity. The purpose of this device is to cause a small amount of current

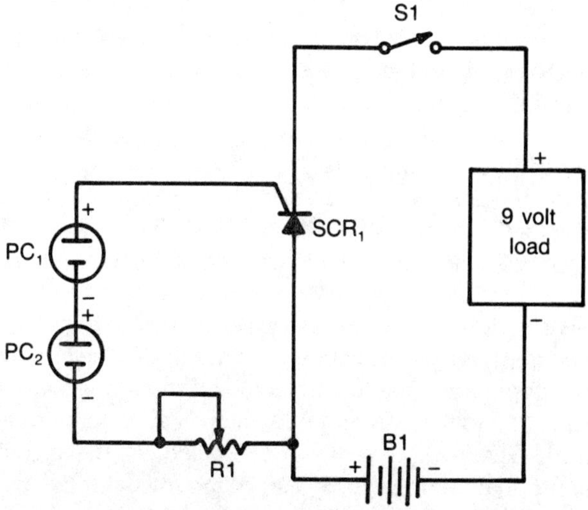

B1—9 volt
PC_1-PC_2—Small solar cells
R1—1 K Ω variable
S1—SPST
SCR_1—SK3950 (RCA)

Fig. 6-13. Solar-trigger switch circuit.

to flow within the gate circuit to cause the device to fire. The resistor simply sets up a separate circuit leg which operates from the same power source. However, a battery could be used here (in the gate circuit alone) to bring about the same triggering action. You can also use any other device which will generate a small electric current.

The circuit in Fig. 6-13 can be thought of as a light-activated SCR switch. Here, a pair of series-wired photovoltaic cells, often known as *solar cells,* is used to provide a separate power source for the SCR gate circuit. A 9-volt battery is used elsewhere in this circuit to supply current to the load.

For this project, I used the RCA SK3950 SCR, which is rated to deliver a maximum of 800 milliamperes to the load. It requires the gate circuit to be triggered by a value of approximately 0.8 volts, delivering a current of approximately 800 microamperes (0.8 milliamperes).

A single photovoltaic cell will deliver at least .4 volt and usually .42 volt or more. A single cell will not be adequate to deliver the required .8 volt, so two are wired in series. In this circuit, the voltage values add, so the total potential at the gate circuit will be equal to two times the voltage in a single cell, or approximately .8 volt. Whenever light strikes the surfaces of the solar cells, the SCR will fire. A variable resistor serves as a sensitivity adjustment.

You can use the tiniest solar cells you can find, as even the smallest will deliver 8 milliamperes or so, which is far more than adequate for this circuit. When the SCR fires, it is no longer dependent upon the gate circuit, so S1 is used to remove power from the load.

This circuit may be constructed without the use of perforated circuit board, and a plastic enclosure is desirable. The photovoltaic cells can be glued to the top surface, and small holes are drilled for connection to the SCR and R1, which is mounted through the front panel. The 9-volt battery is also installed within the enclosure in an appropriate spring clip. S1 can be mounted through the front panel as well. The 9-volt output is accessed by means of a two-contact terminal strip mounted through the rear wall of the enclosure.

To test the circuit, connect a transistor radio or other appropriate load to the terminals, and be sure you observe proper polarity. A reversal can quickly damage delicate solid-state components within the load. Now, activate S1 and shine a bright light on the exposed solar cells. You may have to adjust R1 a bit, but the circuit should fire and deliver the full 9-volt potential (less a half volt or so for the internal resistance of the SCR) to the load. Once power has been delivered, the light may be removed from the solar cells and operation will continue.

Again, use the smallest solar cells possible, as high current capability within the gate circuit is neither warranted nor desired. You can use this same circuit to drive much larger silicon-controlled rectifiers, although it may be necessary to wire as many as four photovoltaic cells in series, depending upon the gate trigger voltage rating of the device (VGT). The maximum voltage you will encounter will be about 1.5 volts, even when using extremely high current devices. Using an SCR with a higher voltage rating, you may derive load power directly from the ac line. If ac is used, however, the circuit will be active only when a constant light source is applied to the photovoltaic cells. This circuit may be used only with dc loads, regardless of the voltage potential.

PROJECT 13: VOICE-ACTUATED SWITCH

A voice-actuated switch is a device which causes a switching action any time your voice (or any other sound for that matter) is picked up by a microphone. These devices have been around for some time and have been used to control the actions of electronic toys and even intruder alarms. The idea is simple. The output from an audio amplifier is used to drive the gate circuit of a thyristor. This is the same method that is used to produce most light organs, like the one presented earlier in this chapter. In light organ circuits, however, the audio amplifier is driven by a turntable, tape player, radio receiver or other such audio device.

The circuit shown in Fig. 6-14 includes a simple audio preamplifier which is transformer-coupled to the gate circuit of a triac. The load for this circuit is a 115-Vac relay, which can be of the latching type if desired. Instead of using a standard microphone element, this circuit uses a common 8-ohm miniature speaker, which should be quite easy to obtain locally. When used in this mode, the speaker is really a dynamic microphone. Many individuals think that a speaker can be used only to output acoustic signals when driven by a varying current, but the reverse is also true. When sound waves strike the surface of the speaker, it produces a small, varying electric current which corresponds to the sound waves. The output from this amplifier is quite low, so you will probably have to stick with a low-current triac, which requires minimum drive to its gate circuit.

The entire circuit can be built on a 6-inch square section of perforated circuit board, but be sure to isolate the 115-volt line from the low-voltage portion of the circuit to prevent catastrophic destruction should a short circuit occur. This size circuit board is adequate to contain all components, including the transformer, triac, and relay. The transistor used is a common type, and many other pnp transistors can be substituted with about the same results. Some units, however, may not give you adequate output.

To test the circuit, connect the probes of an ohmmeter to the output terminals of RY1. Insert the plug in a wall outlet and activate S1. With R3 at the mid-range position, clap your hands near the speaker, and the relay should click on. If it does not, try humming into the speaker while adjusting R3. When this latter component is at the correct setting, the audio input to the speaker will hold the relay shut. When you stop humming, the relay will open again. If you need high-current switching capability, connect the

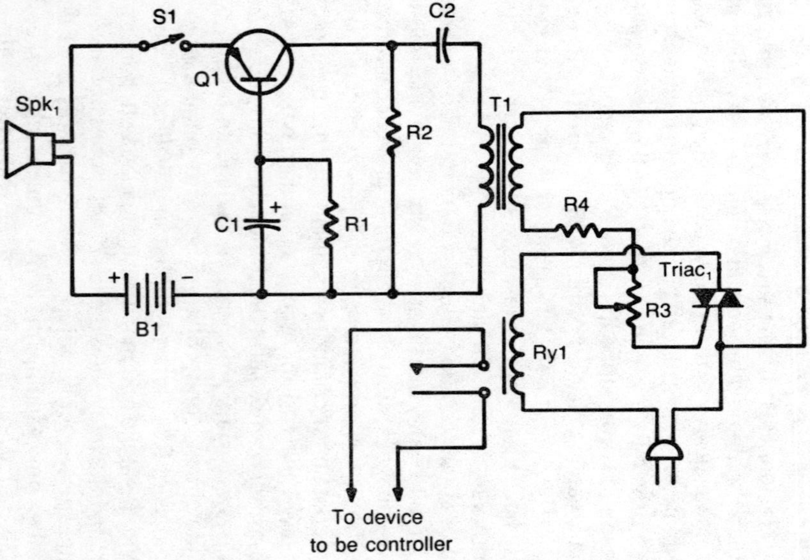

B1—9 volt
C1—10 μF electrolytic 16 Vdc
C2—0.5 μF
Q1—SK3721 (RCA)
R1—250 K Ω 1/2 watt carbon
R2—250 K Ω 1/2 watt carbon
R3—10 K Ω variable
R4—1500 Ω 1/2 watt carbon
Ry1—115 Vac relay
S1—SPST
SPK$_1$—miniature 8 Ω speaker
T1—10 K Ω primary 1000 Ω secondary
TRIAC$_1$—1 ampere 200 PRV

Fig. 6-14. Voice-actuated switch circuit.

power circuit of a larger relay to that of the one specified in the schematic. Relays which are capable of switching 40 amperes or more contain coils which still draw minimal amounts of current and are well within the switching capabilities of even the smallest 115-volt relay. A 1-ampere triac is specified for this circuit, so any 115-Vac relay which draws less than this amount may be safely used, although the larger types cannot be mounted to the circuit board.

For an intruder alarm circuit, use a latching relay which is connected to the power circuit of an appropriate buzzer or other alarm indicator. The device can then be set to maximum sensitivity to respond to any loud sound.

PROJECT 14: SCR CONTINUITY CHECKER

In some test applications, it is difficult to use an ohmmeter due to its size. For this reason, many people who repair electronic equipment and electromechanical devices use a continuity checker to test for short circuits and open circuit conditions. In its most basic form, a continuity checker is often a flashlight battery which is connected to a low-voltage incandescent bulb. The test probes serve as a switch. When there is an extremely low resistance across the probes, the circuit is completed and the bulb lights.

When continuity checkers are used for testing integrated circuits and other small solid-state devices, sometimes the battery voltage is high enough to damage these components. The circuit shown in Fig. 6-15 avoids this by using the low-voltage potential which is present within the gate circuit of a silicon-controlled rectifier. The actual value varies, but this is often less than 1 volt. The load is a 9-volt buzzer which receives current from the 9-volt transistor radio battery. A 1 K potentiometer serves as the sensitivity control, which can be adjusted to provide triggering action when the probes are placed across varying values of low resistance. When a low resistance condition exists at the probes, the SCR fires, conducting current through the 9-volt buzzer. A momentary (normally closed) switch is included in this circuit to reset the device for the next test, as the buzzer will continue to sound even after the probes have been removed. In many test functions, this is a desirable feature. It allows the technician to simply clip the probes in place and make adjustments elsewhere within the circuit. In some devices, these adjustments may be at a point that is quite a distance away from the portion being tested for continuity. When an adjustment causes a circuit to complete itself, the buzzer

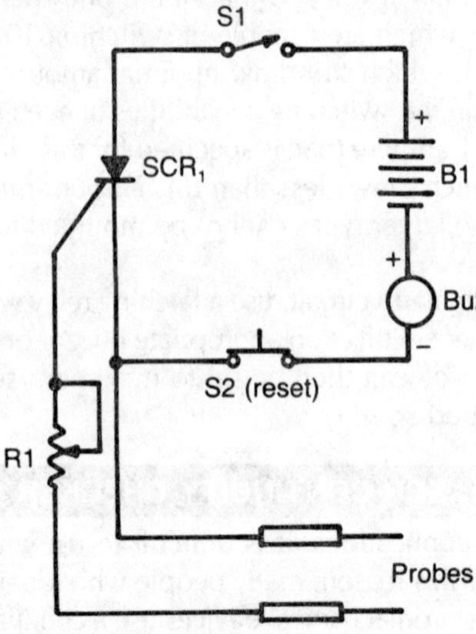

B1—9 volt
Bu2—9 volt buzzer (50 mA or less)
R1—1 K Ω variable
S1—SPST
S2—*normally closed* momentary switch
SCR_1—low current SCR (100 mA or less)

Fig. 6-15. SCR continuity checker circuit.

sounds, and this condition may be remotely sensed.

You may use perfboard for constructing this circuit or forego this altogether and use point-to-point wiring along with terminal strips. A small plastic or aluminum case will serve as a fine enclosure, and this can be fitted with receptacles to accept the plugs from an inexpensive pair of multimeter probes that can be purchased from a local hobby store.

To test the circuit, activate S1 and set R1 to a mid-range position. Short the probe tips together, and adjust R1 until you hear the buzzer. Now, separate the probe tips and press S2 momentarily. This turns the buzzer off. Short the probes together again, and the buzzer should sound. If not, decrease the resistance of R1 slightly, or until positive trigger action occurs each time the probes are shorted together.

This circuit certainly won't replace your ohmmeter, but it does come in handy for a quick check of many different appliances and some electronic circuits. The low-voltage gate circuit has a potential which is probably lower than that of most ohmmeters, so you don't have to worry about damaging delicate electronic components.

PROJECT 15: SCR PROXIMITY ALARM

A proximity alarm is a circuit or device which triggers an indicator whenever a person or object comes close to or actually touches an electrode sensor. Such devices are used for many different purpose, including intruder alarms, counters, and for automation purposes.

The circuit shown in Fig. 6-16 uses a silicon-controlled rectifier which is fired by I1, a miniature neon lamp, based upon circuit capacitance. A miniature 20-picofarad trimmer capacitor is used as the sensitivity control, which is tied in parallel with a sensor plate. The circuit looks incomplete as shown, but a person's body capacitance in relation to the electric wiring in the wall forms the other connecting leg from the sensor. A variable resistance is

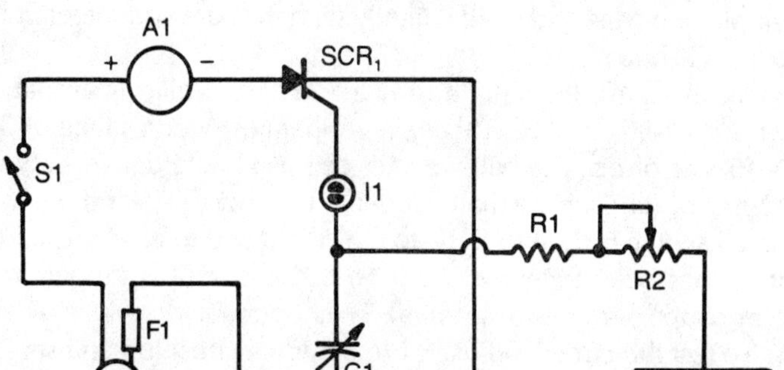

A1—Sonalert (Allied Electronics 854-6500)
F1—1/2 ampere fuse
C1—20 pF trimmer
I1—miniature neon lamp
R1—50 K Ω 1/2 watt carbon
R2—100 K Ω variable
S1—SPST
SCR_1—200 1 ampere

Fig. 6-16. SCR proximity alarm circuit.

included in the sensor lead. In some applications, this will have a reasonably good effect, while in others, it will do little or nothing. This circuit is dependent upon the 115-Vac house wiring, both for supplying power and serving as a means of firing the circuit.

Perforated circuit board is used as a platform for mounting all components. The completed circuit can be housed in a plastic or aluminum enclosure, which provides access for the line cord and the single lead to the metal sensor plate. The silicon-controlled rectifier may be any 200-volt unit, as the circuit draws a very small amount of current when activated.

The sensor plate may be a doorknob (for intrusion alarms), but this small metal surface will most likely require a direct contact to trigger the circuit. For a true proximity alarm, it is necessary to install a much larger metal detecting surface. The easiest way to accomplish this is to use a large section of aluminum foil which can be tacked directly to the wall or even mounted behind it for a hidden installation. The larger the surface area of the sensor plate, the more sensitive your circuit will be to individuals passing by. It will be very difficult to solder aluminum foil, so you can simply crumple one edge and wrap it tightly with the bare conductor from the circuit board.

As an alarm, this circuit uses a Sonalert, which is available from Allied Electronics. This device operates over a range of 30 to 115 Vac or dc. The half-wave circuit used here delivers about 60 Vdc, which is more than adequate to drive this alarm device. A small switch is used to activate the circuit and to reset the alarm once it has been triggered. As soon as the circuit is tripped, the alarm sounds continuously until S1 is opened.

To test the circuit, adjust C1 to a midway position and set R2 to a minimum resistance position. Activate S1 while the detection area is clear. Now, have an assistant enter the detection area, standing close to the metal contact plate. Adjust C1 until the alarm sounds. Then, open S1, and the alarm should stop immediately. Have your assistant exit the detection area and close S1 again. The alarm should remain silent. When the person enters the detection area again, the alarm should sound.

If you have problems getting your circuit to fire, have your assistant actually touch the metal plate while adjusting C1. If this is the only way the circuit will fire, your plate area is not large enough. Once your circuit has been initially set up, you can adjust R2 in an attempt to desensitize the circuit somewhat. In most instances, however, C1 will be adequate for this function and R2 may be eliminated altogether.

PROJECT 16: SCR POWER MONITOR

A power monitor or power failure alarm, as it is sometimes called, is a circuit which indicates that a power failure has occurred, even for a brief instant. This type of alarm is quite useful in critical applications where it is necessary to supply operating power to a device such as a clock or other timing circuit on a continuous basis. If the power should fail for only a second or so and then come back on again, the timing pulses from the clock circuit may be erratic or out of sync. Some power failures associated with the ac line or dc power supply may occur briefly at a time when you are not present. Without a power failure alarm, you probably will not be aware of this condition immediately upon return, and it may take many hours to figure out why a particular timing circuit does not seem to be operating properly. In fact, it is operating properly, but the timing sequence has been interrupted by the brief power failure.

The circuit shown in Fig. 6-17 is quite simple, inexpensive, and will definitely indicate when a power failure has occurred. It acts upon the latching effect of a silicon-controlled rectifier, which is fed dc current. Should the current fail for even a split-second, the SCR switches to the off state and will not be fired again until the trigger circuit is manually activated. This circuit is designed to operate from a dc power supply which is connected to the ac line. The circuit is quite standard, in that when the SCR fires, a small incandescent lamp rated to operate at the power supply voltage glows. Positive triggering is accomplished by adjusting R1. The SCR is fired by pressing S1, which is a normally-off momentary switch that returns to the off position with the aid of an internal spring after it has been initially depressed. After the SCR is fired, it is no longer dependent upon the gate circuit to control conduction, so the gate circuit is electronically removed. This makes the lamp glow as long as there is dc power. But remove it for an instant, and the SCR switches off, extinguishing the lamp. Should power be applied again, the SCR is still cut off because of the break in the gate circuit. The only way the lamp can glow again is by manually depressing S1. As long as the lamp remains lit, you can be certain no power failure has occurred. However, if it is off, you know that there has been a power interruption, even if power is currently present.

This circuit may be installed within a dc power supply, and

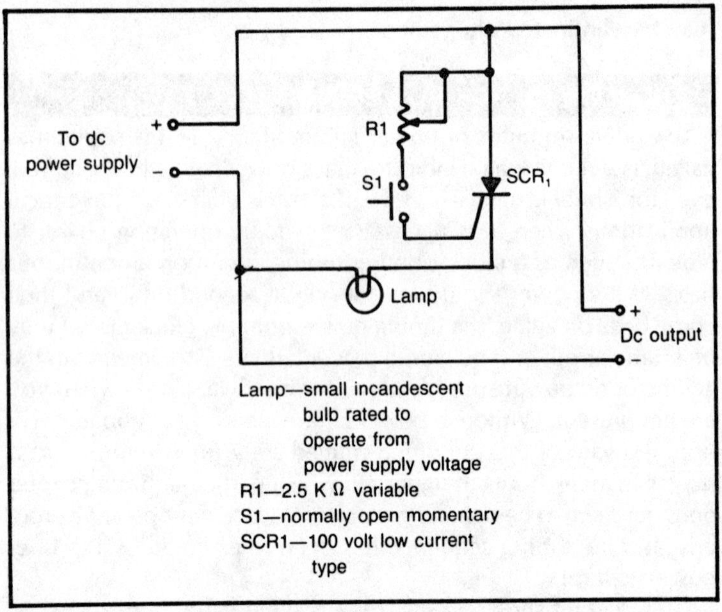

Fig. 6-17. SCR power monitor circuit.

perfboard is not mandatory for construction. R1, S1, and the lamp are mounted through a panel wall, and SCR1 is wired in place to the switch, lamp, and variable resistor contacts. This circuit is wired in parallel with a current dc supply and the electronic load. When the lamp is extinguished, this in no way removes power from the electronic load permanently. When power returns, it is channeled directly to the load, but the lamp circuit will not fire again until S1 is depressed.

To test the circuit, adjust R1 to a mid-range position and activate the dc supply. Depress S1, and if the lamp doesn't glow immediately, adjust R1 while holding S1 closed until it does. Then, remove dc power; the lamp should go out. Apply power again, and the lamp should remain off until you depress S1 again. It should take you only a few minutes to adjust the circuit for positive firing and test it by removing dc power. You may install a normally closed momentary switch between the cathode of the silicon-controlled rectifier and the lamp to be used for testing purposes. This will allow you to extinguish the lamp without actually removing power at the dc supply.

This circuit may be used with any dc power supply which outputs from 5 to about 50 Vdc. The current capability of the supply is not a consideration here, as long as it is adequate to power the electronic load and supply the needed current to light the lamp. Therefore, this circuit can be used with dc supplies that deliver 500 milliamperes or so, on up to massive supplies which may deliver 40 amperes or more. It can not be used with ac supplies, since the constant cycling will cause the lamp to be extinguished each time the sine wave passes through zero. This circuit is quite dependable and can be used for any number of critical electronic applications where it is necessary to supply a constant source of current.

PROJECT 17: SCR SMOKE ALARM

Smoke alarms have become very popular in recent years and are often required by law in newly constructed residences. They remain in the armed state at all times and are triggered only when a small amount of smoke enters a special chamber, which is open to the room it guards. When this occurs, an alarm is triggered and sounds continuously until the circuit is manually reset. These devices have been instrumental in saving many lives and are highly recommended for every home, old and new.

The circuit shown in Fig. 6-18 can serve as a smoke alarm, but this is a hobby circuit and cannot be used in place of a commercial model. Its triggering action is not nearly as dependable, because it doesn't use the special detection chambers found in commercial devices. Most of these are based upon a radioactive element which is not conductive to hobby construction. This circuit may be used as a backup alarm and for experimental purposes, but again, it is not dependable enough to be used as the primary warning of a fire.

The circuit uses a silicon-controlled rectifier whose gate is controlled by a cadmium-sulfide cell. This device was used in a previous project and is a light-dependent resistor whose internal resistance changes with light intensity. In this circuit, the silicon-controlled rectifier is conducting during the armed state. This supplies power to a normally closed dc relay. When the relay is activated, its normally closed contacts are held open and current is not delivered to A1, a 12-volt buzzer or Sonalert, which serves as the alarm. However, when the alarm is triggered by the presence of smoke, the SCR switches off and the relay is deactivated. This allows its normally closed contacts to meet and current is delivered to the buzzer.

This circuit contains its own internal light source, which is provided by a 12-volt incandescent lamp. The lamp should be as small as possible, and certain modifications may be necessary to bring about positive triggering action. The lamp is connected directly across the transformer and is lighted continuously. One of the failings of this alarm lies in the fact that once triggered, it may automatically be reset again should the smoke disappear. This circuit acts upon the light-blocking ability of smoke which enters the space between the lamp and the cadmium-sulfide cell. When there is no smoke present, the photocell has a low-resistance value and causes the SCR to conduct current on each half cycle of alternating current from the transformer secondary. However, when enough smoke is present, the resistance of the photocell rises due to the decreased light level, and the SCR switches off, which completes the circuit through the buzzer. A diode rectifier is used in the buzzer circuit, along with a filter capacitor, to supply pure dc.

The circuit can be installed on perfboard, but you must extend the lamp and PC1 somewhat, as this is the smoke receptor. These components may be installed in a separate plastic or aluminum enclosure, the bottom of which is opened to allow for

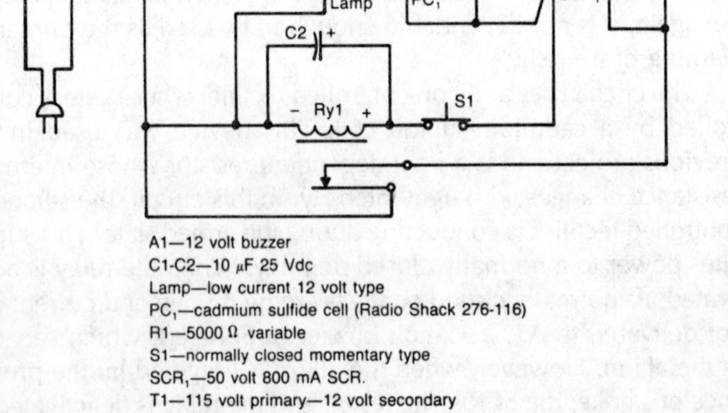

A1—12 volt buzzer
C1-C2—10 μF 25 Vdc
Lamp—low current 12 volt type
PC1—cadmium sulfide cell (Radio Shack 276-116)
R1—5000 Ω variable
S1—normally closed momentary type
SCR1—50 volt 800 mA SCR.
T1—115 volt primary—12 volt secondary

Fig. 6-18. SCR smoke alarm circuit.

the entry of smoke. Variable resistor R1 allows you to adjust the sensitivity of the circuit to various quantities of smoke.

To test the circuit, strike a match to the end of a piece of rope and allow the smoke to flow into the enclosure containing the lamp and PC 1. Then, increase the resistance of R1 until the alarm sounds. If the alarm sounds immediately upon circuit activation, then adjust R1 until the alarm switches off. It may be necessary to apply a small quantity of masking tape to the surface of the lamp to cut down on the amount of light it emits. You may also tape up a portion of the sensitive surface of the photocell. Alternately, you can install a 100-ohm variable resistor in the lead to the lamp to control its output. By carefully adjusting the intensity of the lamp and R1, you should be able to arrive at a setting which will allow the circuit to be triggered in the presence of a small amount of smoke. If you can find a relay which will latch closed when in the off position, this will make the circuit far more reliable, in that the alarm would remain in the triggered state until manually reset. Remember, however, that this is an experimental circuit to show how smoke detectors work, and it is not to be used to replace any of the reliable commercial models which are available almost everywhere for a small price as compared to the life-saving service they provide.

PROJECT 18: BATTERY-TRIGGERED AC SWITCH

Silicon-controlled rectifiers and triacs are often triggered by sampling a portion of the main supply voltage and applying this between the anode and the gate. This is often done with a resistor or variable resistor, the latter being used to control the angle of conduction. However, it is also possible to separate the gate circuit from that of the anode and cathode, using a separate power supply to trigger the gate.

Figure 6-19 shows a method of triggering a triac using direct current while the main part of the circuit operates at a 115-Vac potential. The triac is able to conduct current in two directions and therefore may be used as a control source for alternating current. Here however, two power supplies are used. The first is derived from the ac line, while the second is self-contained in a 9-volt battery and, of course, is direct current. While ac will flow between the two power terminals of the triac, direct current will also flow between one power terminal and the gate. A 9-volt battery is more than adequate to trigger even high current triacs, and in most instances, a 1.5-volt battery could be used just as easily. As

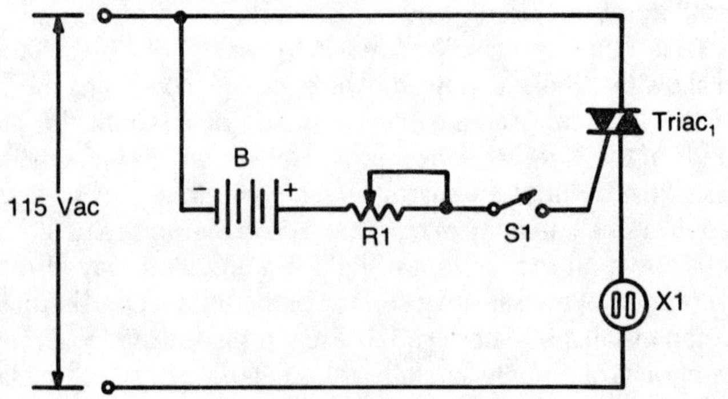

B1—9 volt
R1—200 Ω variable
S1—SPST
Triac—SK 3665
X1—115 Vac receptical

Fig. 6-19. Battery-triggered ac switch circuit.

shown, the triac circuit is an alternating current switch which is controlled by a mechanical switch (S1). This completes the gate circuit. There is no current drain from the battery until S1 is activated. At this time, a milliampere or so will flow to trigger the triac. The 115-Vac supply is delivered to X1, an electrical receptacle, when the triac is triggered. A variable resistor is used to obtain positive triggering action. Typically, its value will have to be adjusted to a resistance of approximately 60 ohms, but this may vary from device to device.

Actually, it would probably be less expensive to simply omit B1 and increase the resistance value of R1 to several thousand ohms and then connect one lead of R1 directly to the top of the circuit where the negative lead of B1 is presently attached. However, this circuit makes a lot more sense when you replace S1 with a miniature relay or even the output from a low voltage dc power source (after removing D1 from this circuit). Here, extraneous circuits which deliver a dc output may be used to switch a high-powered ac circuit. Some computers output a 12-volt potential when certain conditions are sensed, and this power can be used to switch on a high-current ac device. By removing S1 and replacing B1 with a variable dc supply, the value of the ac output at the receptacle could be made to vary with the amount of dc voltage applied. This would allow you to control an ac value with high current potential by means of a low-current dc voltage.

Admittedly, this is a circuit which is designed to be connected to other circuits to form a highly useful function. By itself, it can be used to switch high current ac with a miniature low-current switch. When incorporated in other equipment, however, this circuit has many different functions.

Circuit board construction is not required, and in most cases, it is necessary to use an aluminum enclosure to serve as a heatsink for the triac. Once the triac has been bolted into place, its leads may be used as contact points for the other circuit elements, most of which are mounted through the enclosure wall. The battery is clipped in place, and a terminal strip is used to make the appropriate connections to the ac line and one contact of R1.

To test the circuit, connect an ac voltmeter across the receptacle terminals and adjust R1 to a maximum resistance position. Now, activate S1 and Adjust R1 for a full ac output of 115 Vac nominally. You may also increase the value of R1 to provide a means of adjusting the ac output value. When complete, you will have an excellent ac source which contains no switches or controls within the main alternating current circuit.

PROJECT 19: SCR OVERVOLTAGE PROTECTION CIRCUIT

Some solid-state devices operate over a wide range of dc voltages. Others may operate over an equally wide range, but will output erratic pulses or other products that are voltage-controlled. Still others can be permanently damaged by even a mild deviation in supply voltage. These latter two classes are addressed by this project. In situations where it is necessary to deliver a fairly critical voltage value, it is also necessary to provide a means of overvoltage protection. Such circuits often take the form of automatic control devices which will correct an overvoltage situation. Others will remove power to the electronic load when voltage values rise above a certain preset point. The circuit shown in Fig. 6-20 uses the gate circuit of an SCR to sample the input voltage value. The SCR is installed in parallel with the 12-volt line and includes a normally closed 12-volt relay. The relay contacts are wired in series with the line supplying current to the electronic load. As long as the voltage value at the input remains below a certain point (in this case, 12 volts), the SCR is in its non-conductive state, and RY1 is not activated. Therefore, its normally closed contacts complete the load circuit and power is delivered. However, should voltage rise, the setting of R1 causes enough current to flow within the SCR gate circuit to fire the SCR and thus, deliver current to RY1. When the relay is activated, its contacts snap open and power is removed from the electronic load.

The circuit uses a low current SCR and a relay, which can be mounted on perfboard. Many relays offer multiple contacts, some of which are normally open and others are normally closed. You must use the normally closed contacts here, or the circuit will operate in reverse, delivering power only when an overvoltage condition exists.

The simplest way to construct this circuit is to mount all components on the perfboard and interconnect them as shown. Then, connect the electronic load leads once the circuit is complete. The SCR circuit, including the relay coil, is actually in parallel with

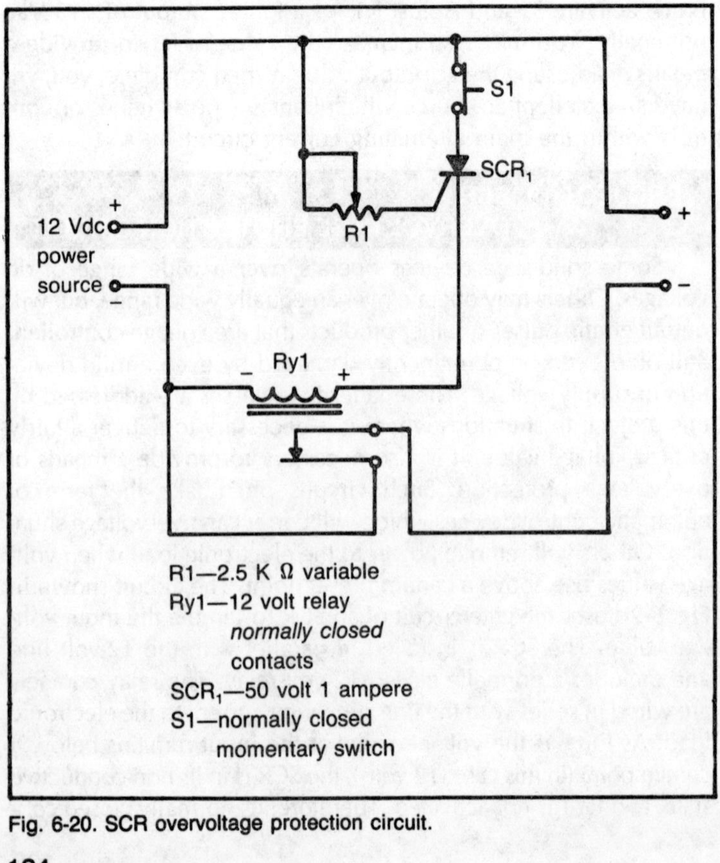

Fig. 6-20. SCR overvoltage protection circuit.

the dc line. The positive lead to the electronic load is connected directly to the positive power supply lead, while the negative load lead is connected to one of the relay contacts. The negative power supply lead is connected to the other.

The circuit can be installed in a plastic or aluminum enclosure or within the enclosure of the existing power supply. A two-terminal contact strip is used for the electronic load outputs.

To test the circuit, connect the input to a dc supply which has a variable voltage option. Connect a highly accurate dc voltmeter to the output terminals. Adjust R1 to its high resistance setting and activate the 12-volt power supply. Now, adjust the supply to a value of just over 12 volts. This can be read directly on the meter. Then, back off on R1 very slowly until you hear the relay click on. Simultaneously, the voltmeter reading should go to zero. Reset the power supply to output 12.0 volts. Press S1 momentarily, and when it returns to its normally closed position, the meter should again read 12 volts. If it reads zero, back off a bit on R1 and depress S1 again. When you get a 12-volt reading at the meter, slowly increase the power supply output voltage. When it reaches a value of 12.5 Vdc, the circuit should again switch off. Use a good variable resistor for this purpose, one which offers smooth adjustment and easily holds its resistance setting.

Once the circuit has been triggered and has removed power from the load, decreasing the power supply voltage will not cause the load power to be reapplied until you depress S1. The power will then be reapplied, assuming the power supply value has dropped below the trigger point. The momentary switch simply removes power from the triggered SCR. This is the only way you can re-arm the circuit.

This is an excellent overvoltage protection circuit because it doesn't require any current from the source to be operated in the armed state. The circuit is constantly on guard and draws current only when voltage exceeds the preset value. In this condition, only the protection circuit is activated. The electronic load is completely removed by the opening of the relay contacts.

While this circuit is designed for 12-volt operation, you may use it for many other power supply values, although it may be necessary to adjust the value of R1 accordingly. Should your dc power supply become defective and deliver an output far in excess of 12 volts, the protection circuit will remove the power from the load and possibly prevent serious and quite expensive electronic damage.

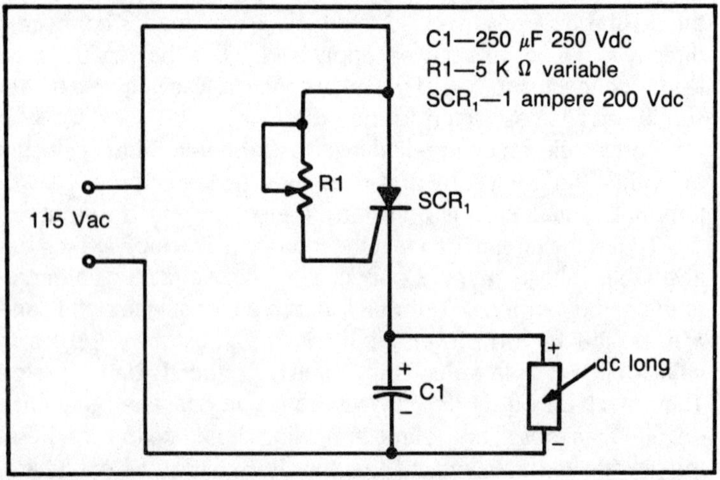

Fig. 6-21. SCR variable dc power supply circuit.

PROJECT 20: VARIABLE DC POWER SUPPLY

When a silicon-controlled rectifier operates from a pure dc current, it serves only as a switch rather than a control of output voltage. When the SCR fires due to the current flow in its gate circuit, the full dc output is delivered to the load and will not be further affected by current flow in the gate circuit. When an SCR operates from an alternating current input, the device is dependent upon its gate circuit to retrigger. The ac sine wave is constantly fluctuating between positive and negative values, passing through zero at the completion of each half cycle. The SCR is still acting like a switch. When it is first triggered to the conducting state, however, it switches off again when the current value approaches zero. Since this occurs 120 times per second, the SCR must be retriggered every 1/120th of a second for a constant output. By controlling the conduction rate of the SCR, we can control the value of output voltage delivered to the load. Therefore, the SCR can be effectively used as a light dimmer, motor speed control, and in many other control applications when fed by an alternating current source. The circuit shown in Fig. 6-21 uses the silicon-controlled rectifier with an ac input to provide a variable dc output voltage. The alternating current is rectified by the SCR, and this pulsating dc output is filtered by an electrolytic capacitor. The setting of R1 determines the output delivered to C1, and thus, the pure dc voltage present at the load.

You can use circuit board construction here, although it is not required due to the small number of components. Connect the positive capacitor terminal to the cathode of the silicon-controlled rectifier. The negative terminal is connected across one side of the ac line and to the negative side of the dc load. The maximum output voltage value from this circuit will be approximately 160 Vdc, so your electrolytic capacitor must be rated for 200 Vdc or more.

To test the circuit, advance R1 to its maximum resistance position and apply ac power. Connect a dc voltmeter across the output and begin to adjust R1. As R1 decreases in value, the voltage will slowly rise. The value read at the meter will remain constant for a second or so when you attempt to lower voltage by increasing the resistance of R1. In other words, the output reading will lag the adjustment of R1 by a short period of time. This is due to the charging effect of C1. When charged to a maximum value of 160 Vdc, the meter will read this amount, even when the pulsating dc output from the SCR is decreased. The capacitor holds this high charge until the load bleeds it off. Again, this occurs for only a second or so (depending on the load resistance), and the circuit shown should deliver dc outputs of from 0 to 160 volts. The variable resistor should be a high quality linear taper variety which can be purchased at most electronic hobby stores. Incidentally, you can also use this circuit with a 230-Vac input by increasing the rating of the silicon-controlled rectifier to 400 volts dc and the rating of C1 to 500 Vdc. You will then have a maximum power output of approximately 320 volts.

This is not an electronically regulated power supply, so the voltage output will vary if the load draws varying amounts of power. You probably won't be able to get away with operating integrated circuits and other solid-state devices with this supply operating in the low voltage output mode due to the high filtering requirements normally specified for such circuits. As a variable dc source, the simplicity of the supply is quite unparalleled.

PROJECT 21: TRIAC-CONTROLLED VOLTAGE DOUBLER

Most electronic power supply circuits operating from the ac line use iron core transformers to step up or step down the 115-Vac input potential. However, it is possible to increase the dc output voltage in steps of two or more by using rectifier and filter combinations acting upon the original ac input. Such circuits are called voltage multipliers and operate when connected to any ac source, which may be the secondary of a transformer or the ac line itself.

Figure 6-22 shows a triac-controlled voltage doubler circuit whose input is ac, while its output is dc. A silicon-controlled rectifier would not work here, because it is necessary to deliver an ac potential to the separate rectifier circuits composed of D1 and D2 and C1 and C2. The latter circuit is a standard voltage doubler which uses two half-wave rectifier filter combinations in series aiding. This delivers a dc output potential which is equivalent to twice the peak ac value. Most household outlets deliver 115 Vac, but this is the rms or average potential. Actually, the ac value will soar to a peak of 160 volts, and this is the value to which the capacitors will charge. Therefore, a voltage doubler circuit will deliver a dc output which is equal to nearly three times the ac line rms value. This is the maximum output for this circuit, which can be measured as approximately 320 Vdc. The triac is used to control conduction and therefore, the value of the ac output. The ac is delivered to the rectifiers and filters, where it is converted to pure dc and delivered to a direct current load.

Looking at the rectifier and filter circuit portions, it can be seen

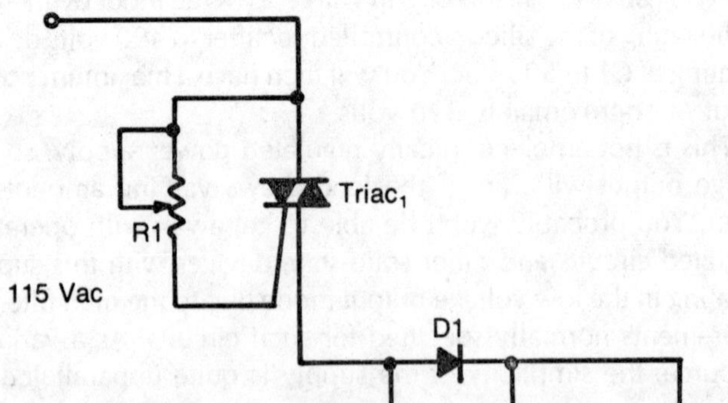

C1-C2—250 μF 250 Vdc
D1-D2—1000 PIV 1 ampere
R1—5 K Ω variable
Triac$_1$—1 ampere 200 volt

Fig. 6-22. Triac-controlled voltage doubler circuit.

that the positive going ac wave is rectified by D1, while it is blocked by D2. The pulsating output from D1 charges C1 to the maximum ac value. During the next half cycle, the polarity swings negative, and the wave is blocked by D1 and passed by D2. This charges C2 to the maximum ac potential. The dc load lies across both C1 and C2 and the latter two components are connected in series, thus adding their charged values. The delivered output is then twice the peak value of the ac input.

In this circuit, the triac is used only to control conduction and can thus deliver variable ac potentials to the rectifiers. This allows you to vary voltage from a low value to a maximum of approximately 320 Vdc.

The previous project also delivers this maximum value, but only when connected to a 230-Vac line. This one increases the dc output from a value which is half of the ac input for the previous circuit. Like the previous project, this one may also be operated from the 230-volt line, but it will be necessary to increase the triac and capacitor ratings. Using this latter power source, over 600 volts will be delivered as the maximum output to the dc load.

This circuit can be constructed in the same manner as the previous project. A 1-ampere triac will not require external heatsinking, but if you want more power, a higher rated triac will be needed, along with the required heatsinking. In a voltage doubler circuit, the voltage value increases, but the current rating must be decreased. A 1-ampere triac can deliver a maximum of only 500 milliamperes at the dc output, so while your voltage value has been doubled at the dc output, total power consumption must remain the same, since the current drain is halved.

Both of these circuits output potentially lethal values, so proper safety procedures are required during alignment and testing. While no fuses are shown in these circuits, an ac line fuse should be installed at the input to the solid-state devices. To make your circuit more usable, a permanently installed dc voltmeter at the output will allow you to constantly measure the potentials to the dc load.

PROJECT 22: TRIAC AC VOLTAGE CONTROL

The previous project used a triac to control the value of the ac voltage present at the rectifier input. We can go one step further by actually taking a step backward and removing the rectifiers and filters. This results in a control circuit having a variable ac output.

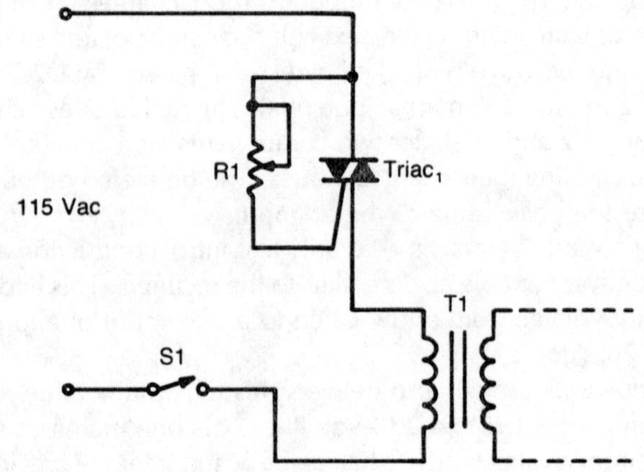

R1—5 K Ω variable
S1—SPST
T1—Existing power supply transformer
Triac$_1$ 200 volt (current rating 20% higher than transformer primary)

Fig. 6-23. Triac ac voltage control circuit.

Figure 6-23 shows this circuit, which is identical to the previous one (without the extra components). This circuit, however, is used to control the output of transformer-operated power supplies which normally derive their inputs from the ac line. It may also be used as a speed control for fans and other ac-operated devices.

This circuit allows you to plug in many different types of devices which contain transformers and to vary their output potentials. For example, when supplying power to an ac-derived dc power supply, you can vary the output voltage in exactly the same manner as in the previous project. This circuit, however, allows you to use existing power supplies and other circuits without internal modifications. You can vary the plate potential to a radio transmitter without accessing the high-voltage secondary.

Early power control circuits of this type used a variable resistor in the ac line. If a high amount of current was drawn by the transformer primary, the variable resistor had to be extremely large, and such units are often quite expensive. They are also quite inefficient, as the large voltage drops they create are converted to heat which must be cast off from the device. This often required external cooling fans. A 10-ampere triac is far less expensive and only a fraction of the size of a variable resistor rated to pass this amount of current.

Of course, high current triac applications will require a large heatsink, but these can be purchased inexpensively. Even with the added size, the control circuit is still smaller and far more efficient than a series-connected variable resistor.

No definite specifications are provided for the triac, other than that it must be rated to handle a 200-Vac potential when operated from the 115-Vac line. The current rating of the triac should be at least 20% higher than (preferably, twice) the maximum current drain at the transformer primary.

To test the circuit, connect an ac voltmeter across the transformer secondary and vary R1. The voltage should move up and down as R1 is adjusted. The transformer is shown here only for discussion purposes and will be a part of an extraneous circuit. In most instances, you will want to install a 115-Vac receptacle for easy access.

Don't use this circuit to control electronically regulated dc power supplies. These devices use solid-state circuitry, which acts upon the dc output from the rectifiers and filters to electronically control the final dc output. If the unregulated dc output from the filter drops below a certain point, the electronic regulator ceases to function properly. Regulated power supplies are usually designed to output a single dc voltage and cannot be controlled at the transformer primary. Variable voltage regulated supplies alter the values of internal electronic components to vary the delivered voltage, and again, are not applicable for operation from this circuit. However, many radio frequency transmitters do not use plate power supplies which are electronically regulated, and this circuit serves as a convenient means of raising and lowering output power by changing the plate potential delivered from the dc supply. Here, control of the primary input voltage works quite nicely.

PROJECT 23: SCR SLAVE FLASH UNIT

For demanding photographic work, it is often impossible to get away with using a single flash unit. It is sometimes necessary to position several flash units at various locations around the subject. Using conventional means, each flash unit must be connected to the camera and triggered by the shutter release. This involves hard wiring which must be run along the floor and hidden from the area which will be captured on film. Invariably, this wiring becomes a source of accidents, as persons moving around the photographic area often trip over them.

An alternative is the use of remote flash units which do not require a direct, hard-wired connection to the camera. These units are light-activated. When the flash unit at the camera goes off, the brilliant light triggers the remote units, which flash in sequence with the master.

Figure 6-24 shows a remote slave flash unit which is controlled by the master flash unit in the camera. This is a self-contained circuit which can be set up and then forgotten. It is designed to operate with commonly available flash bulbs, which must be replaced after each firing. Power is delivered by two 9-volt batteries connected in series to provide an 18-volt source at the flash bulb socket. You can probably get away with a single 9-volt battery here, but the higher potential provides a much more dependable flash action.

The key to this circuit is a light-activated silicon-controlled rectifier (LASCR). When the light from the master unit strikes the sensitized surface of the LASCR, it fires, conducting current through

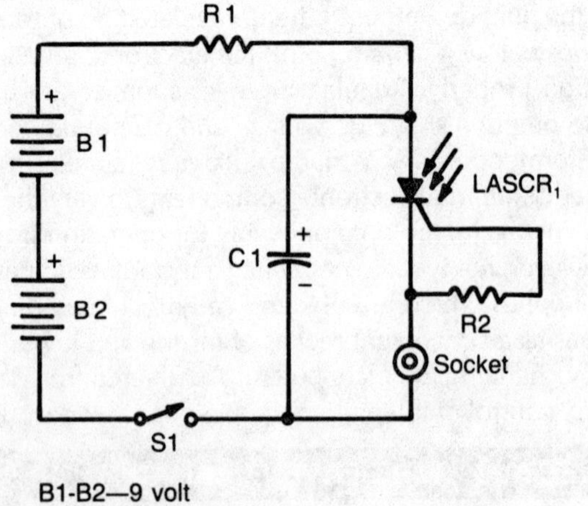

B1-B2—9 volt
C1—360 μF 50 Vdc
R1—10 K Ω 1/2 watt carbon
R2—50 K Ω 1 watt carbon
S1—SPST
LASCR—50 volt 1 ampere

Fig. 6-24. SCR slave flash circuit.

the flash bulb, which is installed within the socket. After the bulb is flashed, its internal element is completely burned out and the circuit is opened, causing the LASCR to cease conduction. When a new bulb is inserted, the circuit is armed again and will be activated upon the next incoming flash from the master unit.

You can install the entire circuit on a piece of perfboard and then mount it in a plastic or aluminum enclosure. The LASCR should be mounted near the edge of the board and a cutout made within the enclosure to allow its sensitized surface to be directed or focused on the master flash unit. A high value electrolytic capacitor (C1) is charged to full potential through R1. The capacitor remains in the charged state, constantly being re-fed by the batteries until the LASCR conducts. When this occurs, the positive and negative terminals of C1 are placed across the flash bulb, and a high energy surge is created which fires the bulb. As soon as the bulb element opens up, C1 is again recharged through R1 and is ready for the next firing of the LASCR. The circuit is deactivated by opening S1, although a full charge will still remain on C1. In other words, there is still a flash charge left in the circuit after the batteries have been completely disconnected.

To test the circuit without using up a lot of flash bulbs, connect a dc voltmeter across the socket terminals. Activate S1 and direct the opening in the enclosure which exposes the LASCR toward the flash unit of your camera. When this flash unit is triggered, you should read a high energy burst at the voltmeter. This indicates that C1 is firing through the LASCR. If you have firing problems, try decreasing the value of R2. Better yet, realign the enclosure opening to intercept more light from the flash unit. After a short adjustment time, you should be able to get positive firing action from this circuit each time the master flash unit is triggered.

PROJECT 24: SCR RELAXATION OSCILLATOR

A relaxation oscillator is an audio type circuit whose operation results from the buildup of a charge in a capacitor, followed by a sudden discharge of a capacitor. The circuit shown in Fig. 6-25 uses a silicon-controlled rectifier in conjunction with a parallel-connected electrolytic capacitor to form a relaxation oscillator which fires through a common 8-ohm speaker. When S1 is closed, the capacitor charges to the full potential of the two 9-volt batteries connected in series for 18 Vdc. Variable resistor R3 is adjusted to cause the SCR to fire. When this occurs, the capacitor is shorted through the speaker, and the sudden discharge

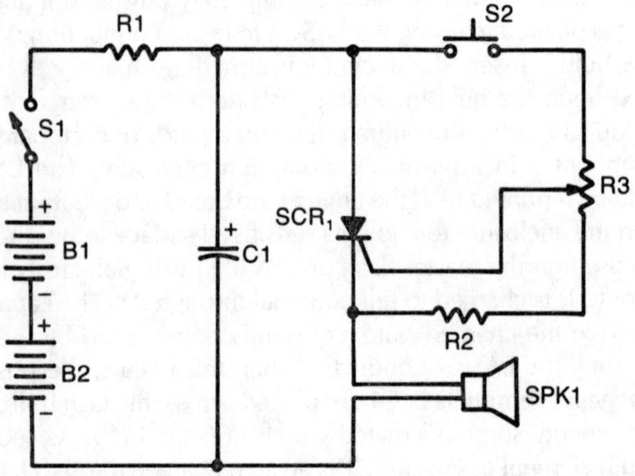

B1-B2—9 volt
C1—1.5 μF 50 Vdc
R1—86 kΩ 1/2 watt carbon
R2—1000 Ω 1/2 watt carbon
R3—1 megohm variable
S1—SPST—S2—Normally open momentary
SCR_1—50 volt 800 mA
SPK1—8 Ω speaker

Fig. 6-25. SCR relaxation oscillator circuit.

creates a raspy note in the speaker. C1 then recharges to a value which is adequate to refire the SCR, and the oscillation continues. This circuit operates upon the continuous charge and discharge of the capacitor. When the capacitor is fully discharged, locking current is removed from the silicon-controlled rectifier, and it switches off for a brief time until the charge value is built back up to an acceptable level. This circuit is fitted with a momentary switch which may be depressed to allow for the SCR to fire initially. As long as the switch is pressed, the circuit will oscillate. When it is released, oscillation stops.

Most people do not think of an SCR in connection with oscillator circuits. However, relaxation oscillators are generally thought of as cheap and dirty circuits, at least when used to produce an audio output. When used as switching oscillators, however, they have many advantages, because they are both simple and highly inexpensive when compared with the use of active solid-state devices.

Assemble the circuit on perfboard, mounting the non-adjustable components here. A plastic enclosure can be used to mount the batteries, S1, S2, R3, and the speaker. There is a current flow anytime S1 is closed, even when there is no output from the speaker. The batteries are used to charge C1, and since this component slowly leaks current, it requires a continuous recharge. When the circuit is not in operation, S1 can be switched to the off position. You can try different values of C1 to produce different frequencies of operation, but the value chosen for this circuit should be about right for most applications. The circuit could be used as an intruder alarm by replacing S2 with a magnetic reed relay, or you can use it as an easily available source of output for aligning audio devices. The output will be of low frequency, and this circuit will sound more like a single piston motorcycle or an old tugboat than it will a pure audio tone. The output volume will not be great, but it will be detectable. For private listening, you can even use a set of headphones connected in place of the speaker. If the circuit does not operate, readjust R3 with the switch held closed. R3 sets the triggering sensitivity of the SCR and must be adjusted to cause the firing action each time S2 is depressed.

SUMMARY

The circuits included in this chapter should be quite interesting to most readers and are eminently practical due to the low cost of components needed. You can probably purchase just a few SCRs and/or triacs and a few capacitors and resistors and build all of the circuits discussed here. By swapping around components in different circuits through the use of plug-in sockets, you might be surprised at what you can build with $10 or $15 worth of components.

Again, many of these circuits operate from the ac line, and this potential must be respected to avoid severe electrical shocks. If a circuit is defective for any reason, be especially cautious, as the full ac potential may be present at a point which would not be active in normal operation. Proceed with caution and always remove the line plug from the wall. If you follow these simple safety rules, your experience in building thyristor circuits will be both rewarding and educational.

Appendix A

Letter Symbols for Signal and Rectifier Diodes

PRV	Peak reverse voltage
I_o	Average rectifier forward current
I_r	Average reverse current
I_{surge}	Peak surge current
V_F	Average forward voltage drop
V_R	dc blocking voltage

Appendix B

Letter Symbols for Zener Diodes

I_F	Forward current
I_Z	Zener current
I_{ZK}	Zener current near breakdown knee
I_{ZM}	Maximum dc zener current (limited by power dissipation)
I_{ZT}	Zener test current
V_f	Forward voltage
V_Z	Nominal zener voltage
Z_Z	Zener impedance
Z_{ZK}	Zener impedance near breakdown knee
Z_{ZT}	Zener impedance at zener test current
I_R	Reverse current
V_R	Reverse test voltage

Appendix C

Letter Symbols for Thyristors and SCRs

I_f	Forward current, r.m.s. value of forward anode current during the "on" state.
$I_{FM(pulse)}$	Repetitive pulse current. Repetitive peak forward anode current after application of gate signal for specified pulse conditions.
$I_{FM(surge)}$	Peak forward surge current. The maximum forward current having a single forward cycle in a 60 Hz. single-phase resistive load system.
I_{FOM}	Peak forward blocking current, gate open. The maximum current through the thyristor when the device is in the "off" state for a stated anode-to-cathode voltage (anode positive) and junction temperature with the gate open.
I_{FXM}	Peak forward blocking current. Same as I_{FOM} except that the gate terminal is returned to the cathode through a stated impedance and/or bias voltage.
I_{GFM}	Peak forward gate current. The maximum instantaneous value of current which may flow between gate and cathode.
I_{GT}	Gate trigger current (continuous dc). The minimum dc gate current required to cause switching from the "off" state at a stated condition.

I_{HO} Holding current. That value of forward anode current below which the controlled rectifier switches from the conducting stated to the forward blocking condition with the gate open, at stated condition.

I_{HX} Holding current (gate connected). The value of forward anode current below which the controlled rectifier switches from the conducting state to the forward blocking condition with the gate terminal returned to the cathode terminal through specified impedance and/or bias voltage.

$P_{F(AV)}$ Average forward power. Average value of power dissipation between anode and cathode.

P_{GFM} Peak gate power. The maximum instantaneous value of gate power dissipation permitted.

I_{ROM} Peak reverse blocking current. The maximum current through the thyristor when the device is in the reverse blocking state (anode negative) for a stated anode-to-cathode voltage and junction temperature with the gate open.

I_{RXM} Peak reverse blocking current. Same as I_{ROM} except that the gate terminal is returned to the cathode through a stated impedance and/or bias voltage.

$P_{GF(AV)}$ Average forward gate power. The value of maximum allowable gate power dissipation averaged over a full cycle.

V_F Forward "on" voltage. The voltage measured between anode and cathode during the "on" condition for specified conditions of anode and temperature.

$V_{F(on)}$ Dynamic forward "on" voltage. The voltage measured between anode and cathode at a specified time after turn-on function has been initiated at stated conditions.

V_{FOM} Peak forward blocking voltage, gate open. The peak repetitive forward voltage which may be applied to the thyristor between anode and cathode (anode positive) with the gate open at stated conditions.

V_{FXM} Peak forward blocking voltage. Same as V_{FOM} except that the gate terminal is returned to the cathode through a stated impedance and/or voltage.

V_{GFM} Peak forward gate voltage. The maximum instantaneous voltage between the gate terminal and the cathode terminal resulting from the flow of forward gate current.

V_{GRM} Peak reverse gate voltage. The maximum instantaneous voltage which may be applied between the gate terminal and the cathode terminal when the junction between the gate region and the adjacent cathode region is reverse biased.

V_{GT} Gate trigger voltage (continuous dc). The dc voltage between the gate and the cathode required to produce the dc gate trigger current.

$V_{ROM(rep)}$ Peak reverse blocking voltage, gate open. The maximum allowable value of reverse voltage (repetitive or continuous dc) which can be applied between anode and cathode (anode negative) with the gate open for stated conditions.

V_{RXM} Peak reverse blocking voltage. Same as V_{ROM} except that the gate terminal is returned to the cathode through a stated impedance and/or bias voltage.

Appendix D

Low-Powered SCR Packages

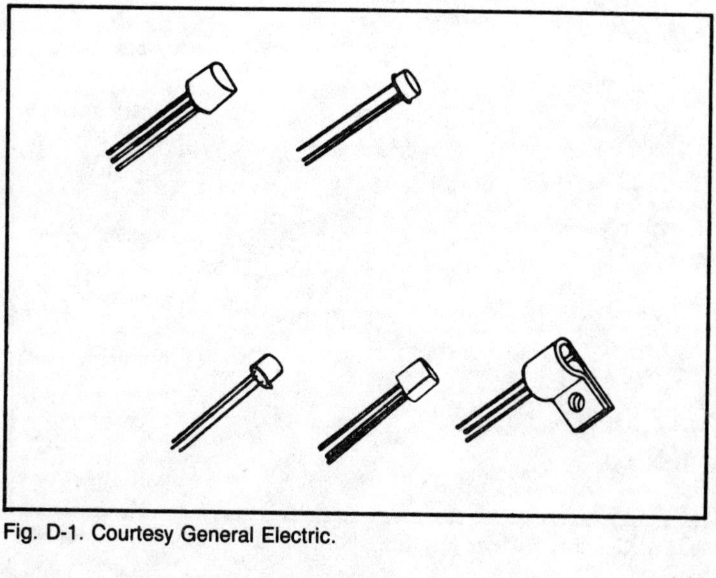

Fig. D-1. Courtesy General Electric.

Appendix E

Moderate to High Power SCR Packages

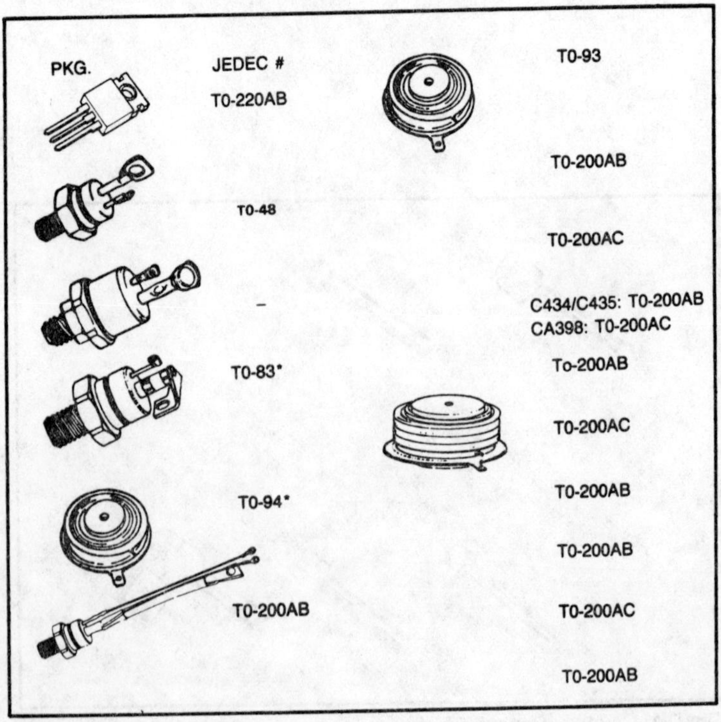

Fig. E-1. Courtesy General Electric.

Appendix F

Triac Packaging

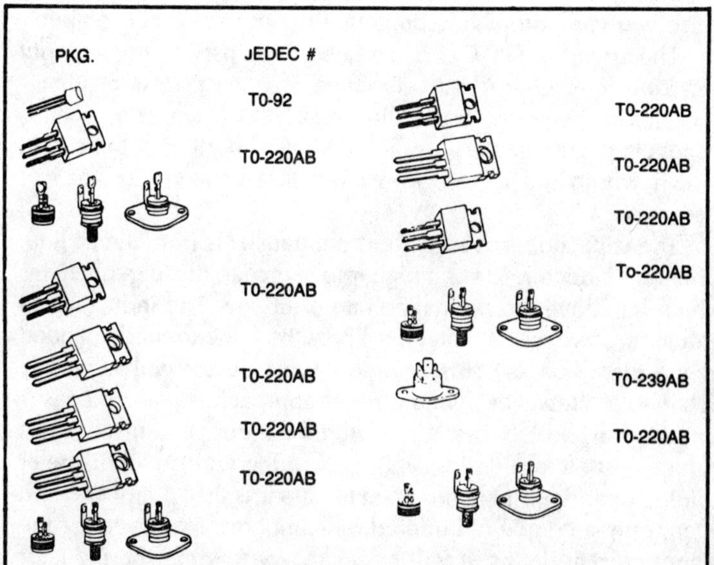

Fig. F-1. Courtesy General Electric.

Appendix G

Thyristor Filter Circuit

During the solid-state switching process, thyristors may create electrical interference which extends into the high frequency portion of the spectrum. When this interference is detected in a receiver or other device with an audio output, it often sounds like a frying noise and can completely obliterate the normal audio output.

The circuit in Fig. G-1 is designed to be used at the ac input of a triac device in order to eliminate or at least substantially reduce this interference. The circuit consists of only two components, a shunt capacitor and a series inductor. The portion of the circuit shown within the dotted lines is all that concerns us for this discussion.

The capacitor is used to short out transients from the ac line, while the inductor acts as a high impedance to those frequencies which lie above the normal ac rate of 60 Hz. The inductor is a problem, however, in that it must pass the full current demanded by the load. For low current drains, the inductor will be readily obtainable. However, when current approaches and exceeds 1 ampere, it may be necessary to wind the inductor yourself. Most inductors used for filtering work are wound with small diameter conductor and most are designed to pass less than 1 ampere, and often, only a couple of hundred milliamperes. However, for the higher current drains, it will be necessary to wind the inductor using No. 18, 16, 14, 12, or even 10 gauge conductor. Obtaining this wire is no problem, but mechanical difficulties can develop.

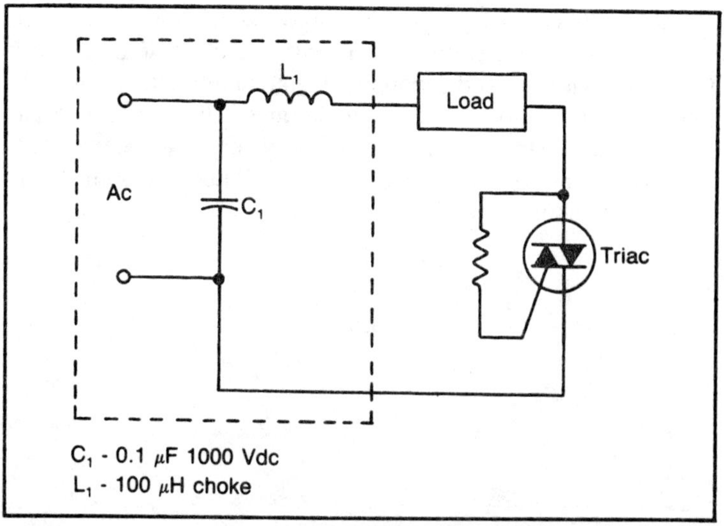

Fig. G-1.

A 100-microhenry choke wound from No. 26 wire will be a small fraction of the size of one wound with No. 12. It is sometimes possible to use smaller diameter wire, but the choke will quickly heat up and waste energy. It also becomes a fire hazard. Appendix H includes a wire chart listing the approved size for varying amounts of current.

In high current applications, you must make certain that all electrical contacts are mechanically stable. This often means that solder is not used, and all connections are made using electrical wiring clamps.

For low current applications, however, a 100-microhenry choke may be found at your local hobby store. Allied Electronics lists a 100-microhenry choke which is rated to conduct up to 2 amperes at a price of less than $3. A 125-microhenry choke rated at 3.5 amperes sells for slightly more. For low current applications, another choke is listed for less than $2. These are often referred to as *hash chokes* and are designed specifically for filtering purposes.

When progressing into the high current range, it may be necessary to convert to an iron core choke, which closely resembles a power transformer, but has only two leads. These may be rated to 10 amperes or more, but they can be expensive when purchased commercially. They are also quite large and heavy. However, many war surplus catalogs list high current iron core chokes for less than $10. The size and weight problems still exist.

Whether or not you decide to install a filter circuit in your thyristor device depends upon the degree of interference and the intensity of operation. If a thyristor circuit is to be used only occasionally, then even a large amount of interference can be tolerated. If, however, it is to be used more or less continuously, some filtering arrangement is mandatory.

Appendix H

Wire Size Chart

Wire Size A.W.G.	Diameter in Mils	Ohms Per 1000 Ft.	Current Carrying Capacity in Amps
1	289.3	0.1264	119.6
2	257.6	0.1593	94.8
3	229.4	0.2009	75.2
4	204.3	0.2533	59.6
5	181.9	0.3195	47.3
6	162.0	0.4028	37.5
7	144.3	0.5080	29.7
8	128.5	0.6405	23.6
9	114.4	0.8077	18.7
10	101.9	1.018	14.8
11	90.7	1.284	11.8
12	80.8	1.619	9.33
13	72.0	2.042	7.40
14	64.1	2.575	5.87
15	57.1	3.247	4.65
16	50.8	4.094	3.69
17	45.3	5.163	2.93
18	40.3	6.510	2.32
19	35.9	8.210	1.84
20	32.0	10.35	1.46
21	28.5	13.05	1.16
22	25.3	16.46	0.918

Wire Size A.W.G.	Diameter in Mils	Ohms Per 1000 Ft.	Current Carrying Capacity in Amps
23	22.6	20.76	0.728
24	20.1	26.17	0.577
25	17.9	33.00	0.458
26	15.9	41.62	0.363
27	14.2	52.48	0.288
28	12.6	66.17	0.228
29	11.3	83.44	0.181
30	10.0	105.2	0.144
31	8.9	132.7	0.114
32	8.0	167.3	0.090
33	7.1	211.0	0.072
34	6.3	266.0	0.057
35	5.6	335	0.045
36	5.0	423	0.036
37	4.5	533	0.028
38	4.0	673	0.022
39	3.5	848	0.018
40	3.1	1,070	0.014

Glossary

anode-to-cathode voltage-current characteristic (anode characteristic): The function, usually represented graphically, relating the principal voltage to the principal current, with gate current (where applicable) as a parameter.

anode-to-cathode voltage (anode voltage): The voltage between the anode terminal and the cathode terminal. It is called positive when the anode potential is higher than the cathode potential, and called negative when the anode potential is lower than the cathode potentials. This term does not apply to bidirectional thyristors.

breakover current ($I_{(BO)}$): The principal current at the breakover point.

breakover point: Any point on the principal voltage-current characteristic for which the differential resistance is zero and where the principal voltage reaches a maximum value.

breakover voltage ($V_{(BO)}$): The principal voltage at the breakover point.

circuit-commutated turn-off time (t_q): The time interval between the instant when the principal current has decreased to zero after external switching of the principal voltage circuit, and the instant when the thyristor is capable of supporting a specified principal voltage without turning on.

critical rate of rise of commutation voltage (for bidirectional thyristor): The minimum value of the rate of rise of principal voltage that may cause switching from the off-state to the on-state immediately following on-state current conduction in the opposite quadrant.

critical rate of rise of off-state voltage: The minimum value of the rate of rise of principal voltage that may cause switching from the off-state to the "on" state.

critical rate of rise of on-state current: The maximum value of the rate of on-state current that a thyristor can withstand without any deleterious effects.

forward current (of a reverse blocking thyristor): The principal current for a positive anode-to-cathode voltage.

forward current of a diode (I_F): The principal current through the diode for a positive anode-to-cathode voltage.

forward surge current of a diode (I_{FSM}): A forward current for a specified waveform and duration.

forward voltage (of a reverse blocking thyristor): A positive anode-to-cathode voltage.

gate controlled delay time (t_{gd}): The time interval between a specified point at the beginning of the gate pulse and the instant when the principal voltage (current) has dropped (risen) to a specified value near its initial value during switching of a thyristor from the off-state to the on-state by a gate pulse.

gate controlled rise time (t_{gr}): The time interval between the instants at which the principal voltage (current) has dropped (risen) from a specified value near its initial value to a specified low (high) value during switching of a thyristor from the off-state to the on-state by a gate pulse. This time interval will be equal to the rise time of the on-state current only for pure resistive loads.

gate controlled turn-on time (t_{gt}): The time interval between a specified point at the beginning of the gate pulse and the instant when the principal voltage (current) has dropped (risen) to a specified low (high) value during switching of a thyristor from off-state to the on-state by a gate pulse.

gate current (I_G): The current that results from the gate voltage. Positive gate current refers to conventional current entering the gate terminal. Negative gate current refers to conventional current leaving the gate terminal.

gate non-trigger current (I_{GD}): The maximum gate current which will not cause the thyristor to switch from the off-state to the on-state.

gate non-trigger voltage (V_{GD}): The maximum gate voltage which will not cause the thyristor to switch from the off-state to the on-state.

gate trigger current (I_{GT}): The minimum gate current required to switch a thyristor from the off-state to the on-state.

gate trigger voltage (V_{GT}): The gate voltage required to produce the gate trigger current.

gate voltage (V_G): The voltage between a gate terminal and a specified main terminal. Gate voltage polarity is referenced to the specified main terminal.

holding current (I_H): The minimum principal current required to maintain the thyristor in the on-state.

I squared t (I^2t): Measure of maximum forward non-recurring overcurrent capability for very short pulse durations. The value is valid only for the pulse duration specified. I is in RMS amperes, and t is pulse duration in seconds. (I^2t is necessary for fuse coordination.)

latching current (I_L): The minimum principal current required to maintain the thyristor in the on-state immediately after switching from the off-state to the on-state has occurred and the triggering signal has been removed.

minimum on-state voltage: The minimum positive principal voltage for which the differential resistance is zero with the gate open-circuited.

mounting force: Range of mounting forces recommended for Press Pak packages to ensure an adequate thermal and electric path while avoiding mechanical damage.

negative differential resistance region: Any portion of the principal voltage-current characteristic in the switching quadrant(s) within which the differential resistance is negative.

non-repetitive peak off-state voltage (V_{DSM}): The maximum instantaneous value of any non-repetitive transient off-state voltage that occurs across the thyristor.

non-repetitive peak reverse voltage (V_{RSM}): The maximum instantaneous value of any non-repetitive transient reverse voltage which occurs across a thyristor.

off-impedance: The differential impedance betwen the terminals through which the principal current flows, when the thyristor is in the off-state at a stated operating point.

off-state: The condition of the thyristor corresponding to the high-resistance, low-current portion of the principal voltage-current characteristic between the origin and the breakover point(s) in the switching quadrant(s).

off-state current (I_D): The principal current when the thyristor is in the off-state.

off-state recovery current (I_{DQM}): The maximum value of the current that results from the application of off-state voltage during the transition from an opposite polarity off-state, a reverse blocking state, or a reverse conducting state.

off-state voltage (V_D): The principal voltage when the thyristor is in the off-state.

on-impedance: The differential impedance between the terminals through which the principal current flows, when the thyristor is in the on-state at a stated opening point.

on-state: The condition of the thyristor corresponding to the low-resistance, low-voltage portion of the principal voltage-current characteristic in the switching quadrant(s).

on-state current (I_T): The principal current when the thyristor is in the on-state.

on-state voltage (V_T): The principal voltage when the thyristor is in the on-state.

peak repetitive on-state current (I_{TM}): The peak value of the on-state current, including all repetitive transient currents.

principal current: A generic term for the current through the collector junction. It is the current through both main terminals.

principal voltage: The voltage between the main terminals. In the case of reverse blocking thyristors, the principal voltage is called positive when the anode potential is higher than the cathode potential, and called negative when the anode potential is lower than the cathode potential. For bidirectional thyristors, the principal voltage is called positive when the potential of main terminal 2 is higher than the potential of main terminal 1.

reapplied rate of rise of voltage, reapplied dv/dt (of reverse block-

ing thyristor): Rate of rise of forward voltage following turn-off or commutation. (This is a test condition for turn-off time measurement.)

repetitive peak off-state voltage (V_{DRM}): The maximum instantaneous value of the off-state voltage that occurs across a thyristor, including all repetitive transient voltages, but excluding all non-repetitive transient voltages.

repetitive peak reverse voltage (V_{RRM}): The maximum instantaneous value of the reverse voltage that occurs across the device, excluding all repetitive and non-repetitive transient voltages.

reverse blocking impedance (of a reverse blocking thyristor): The differential impedance between the two terminals through which the principal current flows, when the thyristor is in the reverse blocking state at a stated operating point.

reverse blocking state (of a reverse blocking thyristor): The condition of a reverse blocking thyristor corresponding to the portion of the anode-to-cathode voltage-current characteristic for reverse currents of lower magnitude than the reverse breakdown current.

reverse breakdown current (I_{BR}): The principal current at the reverse breakdown voltage.

reverse breakdown voltage (V_{BR}): The value of negative anode-to-cathode voltage at which the differential resistance between the anode and cathode terminals changes from a high value to a substantially lower value.

reverse current (I_R): The current for negative anode-to-cathode voltage.

reverse recovery charge (of a reverse blocking triode thyristor or diode) ($Q_{R(REC)}$): The integral of the reverse recovery current.

reverse recovery current (of a reverse blocking triode thyristor or diode) ($I_{R9REC)}$): Negative principal current that flows through the thyristor or diode before it regains its reverse blocking capability.

reverse recovery time (of a reverse blocking thyristor): The time required for the principal current or voltage to recover to a specified value after instantaneous switching from an on-state to a reverse voltage or current.

reverse voltage (V_R): A negative anode-to-cathode voltage.

stud torque: Recommended mounting torque for stud packages.

surge (non-repetitive) on-state current (I_{TSM}): An on-state current of short time duration and specified waveshape.

thermal resistance (of a semiconductor device): The temperature difference between two specified points or regions divided by the power dissipation under conditions of thermal equilibrium.

transient thermal impedance (of a semiconductor device): The change of temperature difference between two specified points or regions at the end of a time interval divided by the step function change in power dissipation at the beginning of the same time interval causing the change of temperature difference.

working peak off-state voltage (V_{DWM}): The maximum instantaneous value of the off-state voltage that occurs across a thyristor, excluding all repetitive and non-repetitive transient voltages.

working peak reverse voltage (V_{RWM}): The maximum instantaneous value of the reverse voltage that occurs across the device, including all repetitive transient voltages, but excluding all non-repetitive transient voltages.

Index

Index

Other Bestsellers From TAB

☐ **BASIC INTEGRATED CIRCUITS—Marks**

With the step-by-step guidance provided by electronics design expert Myles Marks, anyone with basic electronics know-how can design and build almost any type of modern IC device for almost any application imaginable! In fact, this is the ideal sourcebook for every hobbyist or experimenter who wants his to expand digital electronics knowledge and begin exploring the fascinating new possibilities offered by today's IC technology. 432 pp., 319 illus.

Paper $16.95 Book No. 2609
Hard $26.95

☐ **55 EASY-TO-BUILD ELECTRONIC PROJECTS**

If you're looking for a gold mine of exciting, fun-to-build, *and useful* electronics projects that are both easy and inexpensive to put together . . . here's where you'll hit pay dirt!! Here are more than 50 unique electronic projects selected from the best that have been published in *Elementary Electronics* Magazine . . . projects that have been tested and proven! 256 pp., 256 illus. 7″ × 10″.

Paper $14.95 Book No. 1999
Hard $21.95

☐ **THE ENCYCLOPEDIA OF ELECTRONIC CIRCUITS—Graf**

Here's the electronics hobbyist's and technician's dream treasury of analog and digital circuits—nearly 100 circuit categories . . . over 1,200 individual circuits designed for long-lasting applications potential. Adding even more to the value of this resource is the exhaustively thorough index which gives you instant access to exactly the circuits you need each and every time! 768 pp., 1,762 illus. 7″ × 10″.

Paper $29.95 Book No. 1938

☐ **ENCYCLOPEDIA OF ELECTRONICS**

Here are more than 3,000 complete articles covering many more thousands of electronics terms and applications. A must-have resource for anyone involved in any area of electronics or communications practice. From basic electronics terms to state-of-the-art digital electronics theory and applications . . . from microcomputers and laser technology to amateur radio and satellite TV, you can count on finding the information you need! 1,024 pp., 1,300 illus. 8 1/2″ × 11″.

Hard $60.00 Book No. 2000

☐ **THE ILLUSTRATED DICTIONARY OF ELECTRONICS—3rd Edition—Turner & Gibilisco**

The single, most important reference available for electronics hobbyists, students, and professionalist identifies and defines over 27,000 vital electronics terms—more than any other electronics reference published! More than 2,000 new topics have been added to this state-of-the-art 3rd Edition! Every term has been revised and updated to reflect the most current trends, technologies, and usage—with every meaning given for every term! Covers basic electronics electricity, communications, computers, and emerging technologies! Includes nearly 400 essential drawings, diagrams, tables, and charts! It's the only electronics dictionary that accurately and completely identifies the hundreds of abbreviations and acronyms that have become "standard" in the electronics and computer industries! 608 pp. 395 illus. 7″ × 10″.

Paper $21.95 Book No. 1866
Hard $34.95